Lean TRIZ

How to Dramatically Reduce Product-Development Costs with This Innovative Problem-Solving Tool

The **Management Handbooks for Results** Series

Lean TRIZ: How to Dramatically Reduce Product-Development Costs with This Innovative Problem-Solving Tool
H. James Harrington (2017)

Change Management: Manage the Change or It Will Manage You
Frank Voehl and H. James Harrington (2016)

The Lean Management Systems Handbook
Rich Charron, H. James Harrington, Frank Voehl, and Hal Wiggin (2014)

The Lean Six Sigma Black Belt Handbook: Tools and Methods for Process Acceleration
Frank Voehl, H. James Harrington, Chuck Mignosa, and Rich Charron (2013)

The Organizational Master Plan Handbook: A Catalyst for Performance Planning and Results
H. James Harrington and Frank Voehl (2012)

The Organizational Alignment Handbook: A Catalyst for Performance Acceleration
H. James Harrington and Frank Voehl (2011)

Lean TRIZ

How to Dramatically Reduce Product-Development Costs with This Innovative Problem-Solving Tool

By

Dr. H. James Harrington

CEO, Harrington Management Systems

CRC Press is an imprint of the
Taylor & Francis Group, an **informa** business

A PRODUCTIVITY PRESS BOOK

CRC Press
Taylor & Francis Group
6000 Broken Sound Parkway NW, Suite 300
Boca Raton, FL 33487-2742

Printed on acid-free paper
Version Date: 20160915

International Standard Book Number-13: 978-1-138-21677-8 (Hardback)

Visit the Taylor & Francis Web site at
http://www.taylorandfrancis.com

and the CRC Press Web site at
http://www.crcpress.com

I dedicate this book to my loved ones who have passed away: my father, Frank, my mother, Carrie, my friend, Joe, and my wife, Marguerite. There is just no way to fill the void that was left in my life when God decided to take them away.

Contents

Preface

Start your improvement process with some quick wins to get top management's ongoing support.

H. James Harrington

INTRODUCTION

Everyone wants things faster, better, and cheaper—everyone wants more right now at a lower price. Product life cycles have been cut from years to months to days. We used to mail everything, and then airmail became the way to do it if you needed it delivered fast. Then, Federal Express–type mail came along that allows people to put off doing things until the last minute. Currently, we can't even wait to Federal Express it; we send it by email, but even that is too slow, so we need faster and faster Internet services. If it takes an hour to get an email, we are upset. Many people are texting their friends and business associates to give them instantaneous information no matter where they are located. Conversation is becoming a lost art. Everyone wants things to happen faster and faster.

Methodologies like Six Sigma, Lean, supply chain, error proofing, and business process improvement (BPI) are considered the best practices in the current environment, but Henry Ford Sr. was doing all of this and more in the 1910s. Back in the 1910s, Ford cycle time was seven days from iron ore to the completed car. The car was delivered to the dealer within seven days of the time it left the end of the production line. In Ford's support area, they received payment for the car 10 days before they had to pay for the parts that were used to build the car. (Payments were made to the supplier 30 days after the parts were delivered.) Ford required payment on delivery. That is much better than most organizations can currently do.

No—I'm not saying that we should be using Henry Ford Sr.'s 1910 operating procedures in the current environment. If we did, we could only order a new car in *black*. What I am suggesting is that there is a better way to do almost everything we're currently doing and that we have to focus on the true value of everything we do, not on having the latest and greatest. I was told that if I installed a customer relationship management (CRM) system, my customers would be serviced better, and I would get a bigger share of the market. The CRM system did provide better service to our customers, but it did not increase our market share because our competition also installed a CRM system. Modernizing your thinking and operations may not provide your organization with a competitive advantage, but, if you don't, you will be at a significant competitive disadvantage because your competition is updating their facilities and procedures.

Note: In the 1910s, Ford had no Internet, SAP, ERP, or computers to make his Lean systems and processes operate effectively. He used simple, common-sense approaches to design a system that, even at present, can't be beat. We often rely on computers to do our thinking for us, often making the job more complex, when what we should be doing is removing the complexity. Henry David Thoreau offered the best advice that we could give the small, midsize, and large organizations.

> Simplify, simplify, simplify.
>
> **Henry David Thoreau**
> *Author*

Currently, nothing is fast enough. In this time of plenty, everyone wants more and is satisfied with it for shorter periods of time. No one is willing to wait for anything; everyone wants instant contact with the whole world using their computers and cell phones. In the United States, the majority of our children are now equipped with cell phones so that their parents can know where they are every minute. When we don't know where our children are, the satellite tracking system can always locate them. Are these practices working to keep our children safe? No—we have more crime related to children than ever before. Children themselves are trusted less. Arrests of children under 18 are higher than ever before.

In our private lives, fast is a requirement of quality. In our business/professional lives, the same is true. Fast food is the way we eat. Faster and faster trains are competing with air travel. We only have time for short periods of quality time with our children. In Japan, they hire people to visit their elderly mothers and fathers to take the place of their sons and daughters who are too busy.

Total Quality Control (TQC) in the 1950s was a new womb-to-tomb concept, driven by a group of quality professionals. Statistical process control, design of experiments, understanding customer needs, and the excellence of design were all part of TQC.

Total Quality Management (TQM) was the next big thing to come along (1980). It was sold as a long-term culture change. Management was warned not to expect results right away. Every employee was trained on how to work as a team, solve problems, and control their process.

By 1980, Genrich Altshuller's methodologies (theory of inventive problem-solving [TRIZ]) were already developed and in use in some countries to improve the quality of design.

In the 1990s, process redesign focused on streamlining the process by removing waste and the use of current information technology. Typically, process redesign reduced cost, cycle time, scrap, and rework by 30% to 60%. The methodology took 90 days to develop a future-state solution for most processes and, in some cases, even less.

Although the TRIZ methodology was well documented by this point in time, the emphasis in the United States focused on process quality rather than design quality. Toyota's manufacturing process became the benchmark for organizations around the world. This was based upon the mistaken belief that the growth in Toyota was driven by their manufacturing process. The truth of the matter is people buy Toyotas because of their long-term reliability, which is controlled by excellent design rather than the manufacturing process.

The TRIZ methodologies continued to grow during this time, although its acceptance was very low in the United States as it was viewed as a complex approach to designing new or evolutionary products and problem-solving.

One of the major problems we have had with the performance improvement tools, methods, and techniques has been that they didn't produce results fast enough. Everyone wants zero inventory but expects immediate delivery. Long-term planning is six months, and, within five years, it will be measured in weeks.

Management lost its patience with TQC because results were measured in poor quality cost that took years to measure. TQM lost favor because we talked about it taking a long time to change the organization's culture. Process reengineering lost its sponsors because it took up to 12 months to develop a new process design. Process benchmarking slipped back to just a process benchmark (the measure of what best practices are) because it took three to four months to define how to adopt or adapt the best practices of the benchmarking partner's activities to the emotional costs that are related to copying an organization's way of operating. The process redesign methodology takes three months to develop a future-state solution. Six Sigma projects that were supposed to take a maximum of three months are often taking six to nine months to complete. Organizations need a better, faster way to generate new product designs, improve performance, develop processes, and reduce waste. To fill this need, organizations around the world are turning to the Lean TRIZ methodology (LTM). The following definitions are important to understand this methodology:

- TRIZ methodology

  It is *a problem-solving, analysis, and forecasting tool derived from the study of patterns of invention in the global patent literature*. It was developed by the Soviet inventor and science-fiction author Genrich Altshuller and his colleagues, beginning in 1946. In English, the name is typically rendered as *the theory of inventive problem-solving.*
- 40 TRIZ Principles

  These are 40 one- or two-word statements that describe approaches to resolving technical conflict (problems and/or contradictions) that were defined by Genrich Altshuller based upon his study of over 200,000 patents. These 40 TRIZ principles have a twofold purpose:
  - Within each principle resides guidance on how to conceptually or actually change a specific situation or system in order to get rid of a problem.
  - The 40 Principles also train users in analogical thinking, which is to see the principles as a set of patterns of inventions or 40 TRIZ Principles.

  Within each principle resides guidance on how to conceptually or actually change a specific situation or system in order to get rid of a problem. The 40 Principles also train users in analogical thinking, which is to see the principles as a set of patterns of inventions or operators applicable to all fields of study.
- 39 Characteristics of a Technical System

  These are the 39 engineering parameters for expressing technical contradictions defined in the late 1960s.
- TRIZ Contradiction Matrix

  This is a 39 × 39 matrix. The vertical column lists all of the 39 characteristics of a technical system. The horizontal top line also lists the 39 characteristics of a technical system. The horizontal columns refer to typical undesired results. Each matrix cell points to principles that have been most frequently used in patents in order to resolve the contradiction.

- Lean TRIZ methodology (LTM)

  This is an improvement methodology that is designed to bring about rapid improvements/changes to products and processes by defining and implementing the changes that can be quickly identified and easily implemented, thereby reducing the cost and time to bring about improvement and change.
- Lean Process TRIZ (LP-TRIZ)—the Tool

  This is a tool that is used most often to improve processes. It uses a greatly simplified version of the TRIZ contradiction matrix.

The LTM approach is defining solutions in hours, not weeks or months. The secret to the rapid improvement is in the way it approaches these improvement opportunities. In the current hectic, fast-changing, and demanding organizational environment, there are few critical major processes that are eligible for the 90 days required to develop a future-state solution using Six Sigma, process redesign, or benchmarking. But there are hundreds of subprocesses in most organizations that could benefit greatly from a 10% to 20% improvement. This can be done if a team devoted only one or two days to develop the future-state solution or generate the next-generation product faster than the competition. These could also benefit the organization significantly since that solution could be implemented within 30 days for process or 60 days for products. Even evolutionary product designs used to take 6 to 10 months to finalize the design concepts. This is why LTM is the methodology of choice for many organizations. Often, the LTM approach creates a solution that is two to three generations ahead of the present product or service.

> I thought Six Sigma's return on investment (RoI) was good, until I compared its results to LTM results.
>
> **H. James Harrington**

LTM satisfies two important performance improvement needs that every organization faces. They are the following:

1. *Process* design or redesign
2. *Product* design or redesign

The LTM is divided into five phases. They are the following:

1. Phase I: Identifying improvement opportunities
2. Phase II: Preparing for the workshop
3. Phase III: Conducting the workshops
4. Phase IV: Implementing the change (recommendations)
5. Phase V: Measuring results and rewards/recognition

Figure P.1 provides a visual picture of how these five phases are linked together.

> LTM teams often resolve a problem in two days that save a million dollars (RoI of 1,000 to 1 or better).
>
> **H. James Harrington**

Although the LTM consists of five different phases, there are two different paths that need to be taken through the cycle depending upon the root cause of the improvement opportunity. One path focuses on process improvement, and the other path focuses on design improvement. Four of the five phases are basically identical—the major difference occurs in Phase III.

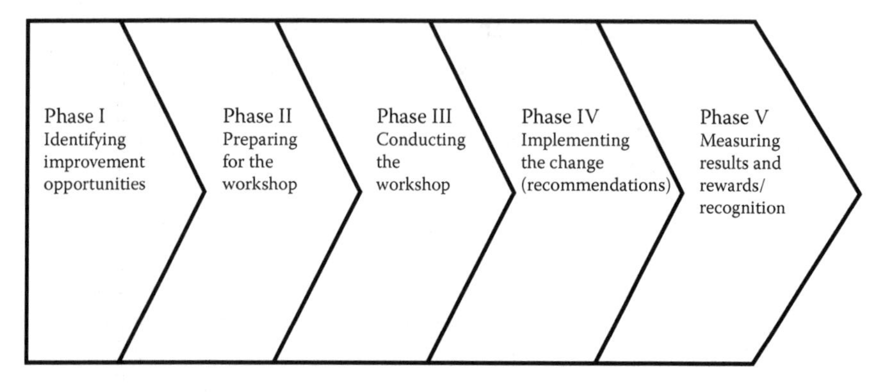

FIGURE P.1
LTM cycle.

LTM WORKSHOPS

Two very different types of workshops are used within LTM due to the very different approaches that are required to improve a design versus improving a process.

- The process design or redesign LTM approach to process-related improvement opportunities focuses upon reducing cost and improving the way the organization develops an evolutionary or new process. This approach centers on a single one- or two-day opportunity development workshop that identifies the root causes of problems, no-value-added activities, and contradictions designed into a present process and/or the proposed process. Process improvement opportunities approach is designed to streamline the processes. In most cases, these are cross-functional improvement opportunities. In these cases, the simple LTM tool—LP-TRIZ—is used:
  - The workshop begins with a brainstorming session.
  - During the next step, streamlined process improvement (SPI) tools are used (e.g., flowcharting, BPI).
  - Next, simple Lean tools are used (e.g., 5Ss, one-minute die change, and just in time).
  - During the final step, a sample version of LP-TRIZ is used to identify contradictions and define ways to eliminate or minimize the impact the contradiction has on the process' performance and/or to reduce the cost of using the process.

During the process design or redesign, LTM workshop teams develop new processes and/or process improvement opportunities.

- The product design or redesign LTM approach is used to upgrade a design or to develop a new product. During this two-day workshop, the team makes use of a number of unique tools.
  - Once the root cause has been agreed to, the team will brainstorm potential solutions for a short period of time.
  - This is followed up with the use of value engineering (VE) methodology.
  - Finally, TRIZ is used to identify contradictions in the proposed improvement and to improve the proposed solution. This will usually reduce the risk of the proposed solution failing to meet

its goals. This often results in a design that is two to three generations ahead of a design that would've been selected without LTM.

Product design or redesign opportunities are all addressed during the product-design LTM process. It is most often used and is most effective when it addresses designs that are initially obvious and/or minor improvements. These two categories account for more than 95% of all the product design activities.

This approach to LTM is used when an engineering design will be needed to take advantage of the opportunity, to design an evolutionary product, or when improvement is required in a single part of the product. In Phase III, a design or redesign improvement workshop is used. Typically, the workshop will start off by reviewing the data that have been collected and agreeing on the root causes related to the improvement opportunity.

The next activity after the product improvement objectives have been set is to conduct a brainstorming session. Not only does this generate some good ideas, but it also encourages the team and helps to stimulate each individual's thinking. A timeboxing approach is used to control the time spent in the brainstorming part of the process.

The next step in this sequence is to conduct a VE study as a way to improve an individual part's performance, designing new or evolutionary products, or to reduce costs in producing a product. (VE methodology was widely accepted and used in the late 1970s and early 1980s.)

In the last step, the team defines the potential ways to improve the item under study by applying TRIZ to the proposed improvement activities. Effectively using TRIZ in this manner minimizes the risks related to project failure and often provides the LTM team with valuable new ideas that often replace the proposed improvements that came out from the VE activities.

Process Improvement Opportunities

Process improvement opportunities make use of many tools that are embedded in formal BPI and SPI methodologies. This includes the following:

- Flowcharting the process.
- Classifying each of the activities as value added, business value added, and no value added.

- Focusing on minimizing the bureaucracy designed into the process.
- Reducing cycle time.
- Eliminating no-value-added activities and minimizing business-value-added activities.
- Minimizing transportation costs.
- Ensuring that the documentation is written in a language that cannot be misunderstood.
- *Negative analysis*—This is an approach where the team creates a list of things that could be done to make the process/product fail. Then, using this list, solutions are developed for all the high- and medium-risk items that were identified during the negative analysis activity.
- *Risk analysis*—The proposed process/product design is then analyzed to identify potential risks that could keep the process/product from functioning as planned. The team then analyzes this list to classify each of the risks as high-, medium-, or low-level risk. Based upon this analysis, plans are put in place to minimize the impact that the high- and medium-level risks will have upon the product, the process, and the organization if they occurred.
- TRIZ is used to maximize the positive impact that the proposed changes will have on the organization. Using TRIZ on the improvement approach helps to identify strengths and weaknesses in the proposed process/product design and to identify major contradictions that are built in the proposed process/product design. It also helps to reduce the risks related to implementing the new or modified design. TRIZ presents an effective methodology that offsets the contradictions. It's also used to provide potential methods to offset the contradictions that are embedded into the proposed design.

LTM uses the simplest yet most effective tools embodied in the TRIZ methodology. It makes use of a modified and simplified version of the TRIZ contradiction matrix, which is designed to solve product problems. It does not make use of the more complex TRIZ tools that are required to solve very complex problems and to come out with breakthrough designs. Typical improvement that results from applying the LTM approach to a process is reducing cost, cycle time, and error rates between 5% and 15% in a 30-day period. When applied to a product or service, it results in the development of a superior design in a two-day LTM design improvement workshop. The potential improvements are identified, and management approves or disapproves the proposed improvement for implementation. The LTM program is designed so that management reviews the teams' suggested improvements, and then they make a decision during the meeting to accept or reject the activity. They must take a position; they are not allowed to put off making a decision. However, in rare cases, the sponsor may request up to five days to make a decision.

An LTM evolves through the following steps:

- A problem, a product, or a process is identified as a candidate for LTM.
- A high-level sponsor agrees to support the LTM initiative related to the process or product that will be improved. (The process or product must be under the sponsor's span of control.)
- An LTM facilitator is assigned, relevant data are collected, and a set of objectives is prepared and approved by the sponsor.
- The team will attend a one- or two-day workshop to develop plans to meet the objectives and develop an action plan that will define a new process design or the new product design. For very difficult and complex problems, we schedule a three- to four-day LTM meeting. This additional time allows us to use many of the more complex TRIZ methodologies. All recommendations must be within the span of control of the team members and be able to be completely implemented within a 30-day time period for a process improvement and 60-day time period for a design of an evolutionary product, component, or assembly. At the end of the second day, the team's recommendations will be presented to the sponsor. Then, the sponsor will either accept or reject the individual recommended improvement. This decision must be made during the two-day workshop. When the team identifies potential improvement opportunities that are outside their span of control, they are presented to the sponsor for consideration at a later date.

The following are ground rules for the workshops:

- The LTM team must agree to be responsible for implementing each recommendation that will be submitted to the sponsor.
- The participants must not be interrupted during the workshop unless it is an emergency personal situation. During the meeting,

cell phones, laptops, and personal computers are turned off. The participants can use them during the breaks.
- At the end of the two-day meeting, the sponsor attends a meeting at which the LTM team presents its findings.
- Before the end of the meeting, the sponsor either approves or rejects the recommendations. It is very important that the sponsor not delay making decisions related to the suggestions, or the LTM approach will soon become ineffective.

Solutions that are approved for implementation are assigned to an LTM team member to implement over the next 30 to 60 days. Each approved improvement is assigned to a member of the LTM who attended the workshop. Efforts should be made to divide the workload among as many of the workshop attendees as possible. The leader of the workshop should follow up to ensure assignments are completed successfully. All too often, the results of an improvement process are estimated, but they are not implemented. Most of them are not an actual measurement of how much the improvement benefits the organization and its stakeholders. Measurement is absolutely crucial to ensure that the work efforts are worthwhile. Typical measurements are the following:

- Increased level of customer satisfaction
- Decreased cycle time
- Increased value per person
- Increased market share
- Increased reliability
- Decreased customer complaints
- Increased profits
- Decreased cost
- RoI

Figure P.2 divides the five phases of LTM into three major headings, along with a short description of the major activities in each of these major blocks. To simplify the figure, we combined the process and design improvement activities into one block.

- Prework
- Workshop
- Implementation

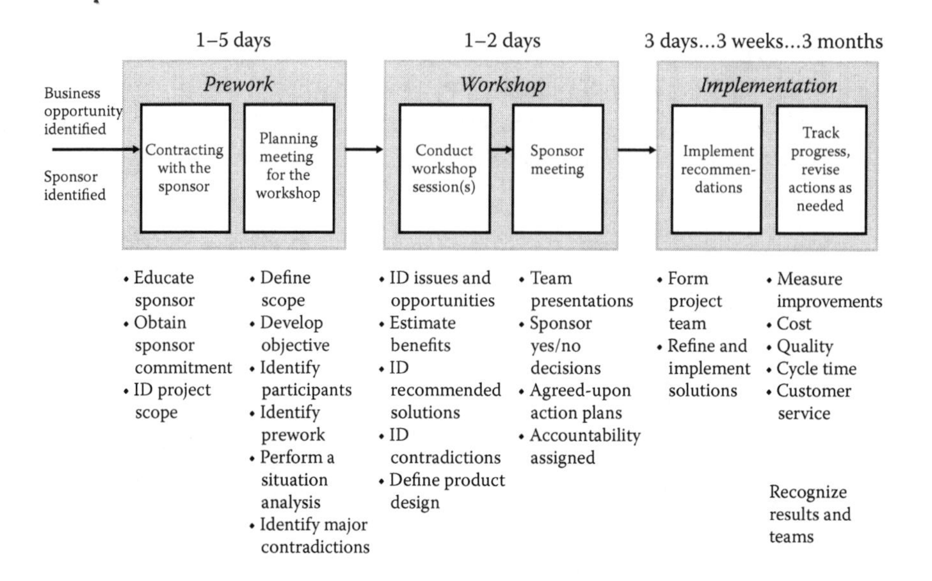

FIGURE P.2
LTM cycle divided into three major headings.

Acknowledgments

I want to acknowledge Candy Rogers, who converted and edited endless hours of dictation and misspelled words into the finished product. I couldn't have done it without her help and proofreading.

I also want to thank my many friends from the Altshuller Institute who worked with me to develop and refine this methodology. I would like to acknowledge all the help and guidance I have received over the years from my two close friends—Chuck Mignosa and Frank Voehl.

About the Author

Dr. H. James Harrington is one of the world's quality system gurus with more than 60 years of experience. In the book, *Tech Trending* (2001), Dr. Harrington was referred to as "the quintessential tech trender." The *New York Times* referred to him as having a "…knack for synthesis and an open mind about packaging his knowledge and experience in new ways—characteristics that may matter more as prerequisites for new-economy success than technical wizardry…." He has been involved in developing quality management systems in Europe, South America, North America, the Middle East, Africa, and Asia.

Present Responsibilities:

Dr. H. James Harrington now serves as the chief executive officer for Harrington Management Systems. In addition, he serves as the president of the Altshuller Institute. He also serves as the chairman of the board for a number of businesses and as the US chairman of chair on technologies for project management at the University of Quebec in Montreal, Canada. Dr. Harrington is recognized as one of the world leaders in applying performance improvement methodologies to business processes.

Previous Experience:

In February 2002, Dr. Harrington retired as the chief operating officer of Systemcorp A.L.G., the leading supplier of knowledge management and project management software solutions. Prior to this, he served as a principal and one of the leaders in the process innovation group at Ernst & Young. Dr. Harrington was with IBM for over 30 years as a senior engineer and a project manager.

Dr. Harrington is a former chairman and former president of the prestigious International Academy for Quality and of the American Society for Quality Control. He is also an active member of the Global Knowledge Economics Council.

Credentials:

The Harrington/Ishikawa Medal, presented yearly by the Asian-Pacific Quality Organization, was named after Dr. Harrington to recognize his many contributions to the region. In 1997, the Quebec Society for Quality named their quality award *The Harrington/Neron Medal* honoring Dr. Harrington for his many contributions to the quality movement in Canada. In 2000, the Sri Lanka National Quality Award was named after him. The Middle East and Europe Best Quality Thesis Award was named "The Harrington Best TQM Thesis Award." The University of Sudan has established a *Harrington Excellence Chair* to study methodologies to improve organizational performance. The Chinese government presented him with the Magnolia Award for his major contributions to improving the quality of Chinese products.

Dr. Harrington's contributions to performance improvement around the world have brought him many honors and awards, including the Edwards Medal, the Lancaster Medal, American Society for Quality's Distinguished Service Medal, and many others. He was appointed the honorary advisor to the China Quality Control Association, and he was elected to the Singapore Productivity Hall of Fame in 1990. He has been named a lifetime honorary president of the Asia Pacific Quality Organization and an honorary director of the Association Chilean de Control de Calidad.

Dr. Harrington has been elected a fellow of the British Quality Control Organization and the American Society for Quality Control. He was also elected an honorary member of the quality societies in Taiwan, Argentina, Brazil, Colombia, and Singapore. He is also listed in the "Who's Who Worldwide" and "Men of Distinction Worldwide." He has presented hundreds of papers on performance improvement and organizational management structure at the local, state, national, and international levels.

Dr. Harrington has two doctor of philosophy degrees—one in quality engineering and an honorary PhD in quality management.

Dr. Harrington is a very prolific author, publishing hundreds of technical reports and magazine articles. He has authored or coauthored over 55 books and 10 software packages. His email address is hjh@svinet.com.

Books by H. James Harrington

The following is a list of some of the many books that Dr. Harrington has authored and coauthored:

- *The Improvement Process* (1987)—one of 1987's best-selling business books
- *Poor-Quality Cost* (1987)
- *Excellence—The IBM Way* (1988)
- *The Quality/Profit Connection* (1988)
- *Business Process Improvement* (1991)—the first book on process redesign
- *The Mouse Story* (1991)
- *Of Tails and Teams* (1994)
- *Total Improvement Management* (1995)
- *High Performance Benchmarking* (1996)
- *The Complete Benchmarking Workbook* (1996)
- *ISO 9000 and Beyond* (1996)
- *The Business Process Improvement Workbook* (1997)
- *The Creativity Toolkit—Provoking Creativity in Individuals and Organizations* (1998)
- *Statistical Analysis Simplified—The Easy-to-Understand Guide to SPC and Data Analysis* (1998)
- *Area Activity Analysis—Aligning Work Activities and Measurements to Enhance Business Performance* (1998)
- *ISO 9000 Quality Management System Design: Optimal Design Rules for Documentation, Implementation, and System Effectiveness (ISO 9000 Quality Management System Design)*—coauthor (1998)
- *Reliability Simplified—Going Beyond Quality to Keep Customers for Life* (1999)
- *ISO 14000 Implementation—Upgrading Your EMS Effectively* (1999)
- *Performance Improvement Methods—Fighting the War on Waste* (1999)
- *Simulation Modeling Methods—An Interactive Guide to Results-Based Decision Making* (2000)

- *Project Change Management—Applying Change Management to Improvement Projects* (2000)
- *E-Business Project Manager* (2002)
- *Process Management Excellence—The Art of Excelling in Process Management* (2005)
- *Project Management Excellence—The Art of Excelling in Project Management* (2005)
- *Change Management Excellence—The Art of Excelling in Change Management* (2005)
- *Knowledge Management Excellence—The Art of Excelling in Knowledge Management* (2005)
- *Resource Management Excellence—The Art of Excelling in Resource Management* (2005)
- *Six Sigma Statistics Simplified* (2006)
- *Improving Healthcare Quality and Cost with Six Sigma* (2006)
- *Making Teams Hum* (2007)
- *Advanced Performance Improvement Approaches: Waging the War on Waste II* (2007)
- *Six Sigma Green Belt Workbook* (2008)
- *Six Sigma Yellow Belt Workbook* (2008)
- *(FAST) Fast Action Solution Teams* (2008)
- *Strategic Performance Improvement Approaches: Waging the War on Waste III* (2008)
- *Corporate Governance: From Small to Mid-Sized Organizations* (2009)
- *Streamlined Process Improvement* (2011)
- *The Organizational Alignment Handbook: A Catalyst for Performance Acceleration* (2011)
- *The Organizational Master Plan Handbook: A Catalyst for Performance Planning and Results* (2012)
- *Performance Accelerated Management (PAM): Rapid Improvement to Your Key Performance Drivers* (2013)
- *Closing the Communication Gap: An Effective Method for Achieving Desired Results* (2013)
- *Lean Six Sigma Black Belt Handbook: Tools and Methods for Process Acceleration* (2013)
- *Lean Management Systems Handbook* (2014)
- *Maximizing Value Propositions to Increase Project Success Rates* (2014)

- *Making the Case for Change: Using Effective Business Cases to Minimize Project and Innovation Failures* (2014)
- *Techniques and Sample Outputs That Drive Business Excellence* (2015)
- *Effective Portfolio Management Systems* (2015)
- *Change Management: Manage the Change or It Will Manage You* (2016)
- *Innovation Tools Handbook, Volume 1: Organizational and Operational Tools, Methods and Methodologies That Every Innovator Must Know* (2016) (Editor)

About the Series

THE MANAGEMENT FOR RESULTS HANDBOOK™ SERIES

As Series Editors, we at CRC/Productivity Press have been privileged to contribute to the convergence of philosophy and the underlying principles of Management for Results, leading to a common set of assumptions. One of the most important deals with the challenges facing the transformation of the organization and suggests that managing for results can have a significant role in increasing and improving performance and strategic thinking, by drawing such experiences and insights from all parts of the organization and making them available to points of strategic management decision and action. As John Quincy Adams once said, "If your actions inspire others to dream more, learn more, do more, and become more, you are a Leader."

If a good leader's actions inspire people to dream more, learn more, do more, and become more (and those actions will lead to an organization's culture, and if the culture represents "the way we do things around here"), then the Management for Results Series represents a brief glimpse of "the shadow of the leader-manager." The Series is a compilation of conceptual management framework and literature review on the latest concepts in management thinking, especially in the areas of accelerating performance and achieving rapid and long-lasting results. It examines some of the more recent as well as historical contributions, and identifies a number of key elements involved. Further analysis determines a number of situations that can improve the results-oriented thinking capability in managers, and the various Handbooks consider whether organizations can successfully adopt their content and conclusions to develop their managers and improve the business.

This is a particularly exciting and turbulent time in the field of management, both domestically and globally, and change may be viewed as either an opportunity or a threat. As such, the principles and practices of Management for Results can aid in this transformation or (by flawed implementation approaches) can bring an organization to its knees. This Management for Results Series (and the Handbooks contained therein) discusses the relationship among management thinking, results orientation, management planning, and emergent strategy, and suggests that management thinking needs to be compressed and accelerated, as it is essential in making these relationships more appropriate and effective—a so-called "shadow of the leader-manager." As Series Editors, we believe that the greater the sum total of management thinking and thinkers in the organization, the more readily and effectively it can respond to and take advantage of the vast array of changes occurring in today's business environment. However, despite the significant levels of delayering and flattening of structures that have taken place in the last decade or so, some organizational barriers continue to stifle opportunities for accelerating management for results by limiting the flow of experiences and insights to relevant corners of the organization.

The "shadow of the leader-manager" that is present throughout this Management Handbook Series is based upon the following eight characteristics of an effective *leader-manager who gets results*, and provides one of the many integration frameworks around which this Series is based:

Integrity = the integration of outward actions and inner values
Dedication = spending whatever time or energy necessary to accomplish the task at hand, thereby leading by example and inspiring others
Magnanimity = giving credit where it is due
Humility = acknowledging they are no better or worse than other members of the team
Openness = being able to listen to new ideas, even if they do not conform to the usual way of thinking
Creativity = the ability to think differently, to get outside of the box that constrains solutions
Fairness = dealing with others consistently and justly
Assertiveness = clearly stating what is expected so that there will be no misunderstandings and dealing with poor performance

This management book series is intended to help you take a step back and look at your team or organization's culture to clearly see the reflection of your leader-manager style. The reflection you see may be a difficult thing for you to handle, but do not respond by trying to defend or to rationalize it as something not being of your making. As difficult as it may be, managers need to face the reality that their team and organization's culture is a reflection of their leadership, leading to the concept of the *leader-manager*. Accepting this responsibility is the first step to change, and as we all know, change begins with ourselves.

As Gandhi said many years ago, we all need to strive to become the "change we want to see in the world".... In the case of *management for results*, we need to be the change we want to see in others!

Frank Voehl and H. James Harrington
Series Editors

1

Introduction to LTM

We live in a world where it seems like everything was needed yesterday.

H. James Harrington

INTRODUCTION

Definition: Lean TRIZ methodology (LTM) is an improvement methodology that is designed to bring about rapid improvements/changes to products and processes by defining and implementing the changes that can be quickly identified and easily implemented, thereby reducing the cost and time to bring about improvement and change.

LTM is a workshop-based process that brings together teams to focus on specific processes, evolutionary product designs, and improvement opportunities. During the workshop, issues are identified, recommendations and action plans are developed, and unfiltered feedback is provided to the project LTM sponsors. The sponsors decide immediately whether or not the recommendations should be implemented.

LTM is also effectively used in designing new evolutionary products based upon a current product. It is sometimes called *picking the low-hanging fruit.* (See Figure 1.1.)

At long last, there is a quick fix that really works. LTM is a magic pill that produces a superior evolutionary product and fixes your business problems without taking the time required to change the organization's

FIGURE 1.1
Picking the low-hanging fruit.

culture. Just as penicillin changed the way doctors treat pneumonia, LTM will give you the same type of miracle results within your organization. It is quick, easy, and painless. Now this may sound too good to be true, but LTM can do it for you.

LTM has been producing solutions in as little as one day and, at the most, two days. It doesn't require four months of training, such as is necessary for a Six Sigma Black Belt. It does not require top management to be trained, nor does it require directing their time to the initiative in order to make it work. After all, top management is already very busy doing strategic planning, defining new products, hiring the right people, and just running the day-to-day business.

As we discussed earlier, there are basically two types of LTM. They are as follows:

1. *Process design or redesign LTM is directed at a specific process or subprocess that is not functioning at the level that the organization would like.* It uses the people who live within the process with the help of one or two facilitators. The future-state solutions are implemented within 30 days by the people who suggested the improvements.
2. *Product design or redesign LTM is based upon taking a current product that is near the end of its life cycle and redesigning it to extend its product life cycle.* The project stays in line with the already-established development, production, and marketing procedures once the basic design is completed using the LTM.

FIVE LEVELS OF INVENTION

There are many ways to improve processes and develop new products. Based upon Genrich Altshuller's studies of more than 200,000 patents and technological systems, he identified that five levels of invention exist. (See Figure 1.2.) They are the following:

1. Level 1—Apparent solutions
2. Level 2—Minor improvements
3. Level 3—Major improvements
4. Level 4—New paradigms
5. Level 5—Discovery

Definition: Five degrees of design complexity is a way of grouping new inventions into five categories that reflect the complexity of the thought pattern that goes into each of the five categories. (See Figure 1.2.)

- Level 1 inventions are obvious and apparent solutions involving well-known methods and knowledge requiring no new invention of any consequence.
- Level 2 inventions constitute minor nonobvious improvements to a system, using methods known within the domain of discourse but applied in a new way.

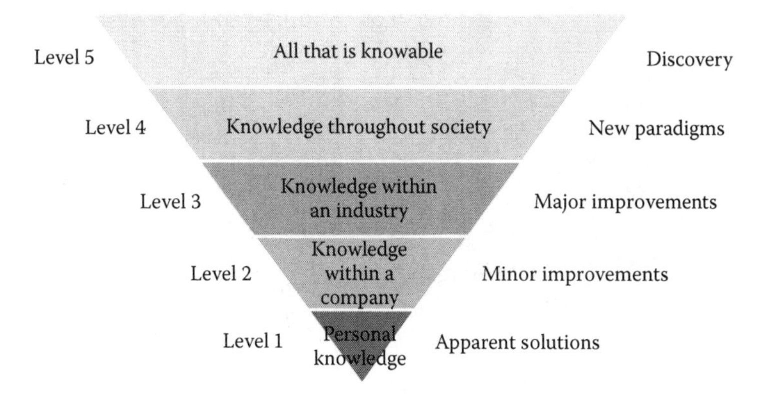

FIGURE 1.2
Five degrees of design complexity.

- Level 3 inventions include fundamental improvements to a system involving methods known outside of the domain. This involves applying an idea to the domain that has never been used in the domain previously.
- Level 4 inventions entail the development of an entirely new operating principle and represent radical changes.
- Level 5 inventions represent a rare scientific discovery or the pioneering of a totally new industry altogether.

After analyzing thousands of patents, Genrich Altshuller established the following curve that indicates the percent of inventions that were recorded based upon the five levels of complexity. (See Figure 1.3.)

Traditional TRIZ focuses primarily on Levels 3 and 4 with some minor impact on Level 5. LTM is designed to assist in generating ideas in Levels 1, 2, and 3. You will note that Levels 1 and 2 account for over 95% of the improvements. As you will see, LTM is designed to assist teams to come to a consensus related to selecting the best design or problem solution for less complex situations. As a result, the LTM team is not required to have a thorough knowledge of the TRIZ methodology. This allows the facilitator to lead the group to reach the best decision in a minimum amount of time.

LTM is not a training or educational process. The main purpose of LTM is to develop a future-state solution that is approved by management during a two day workshop. All too often, approaches like Six Sigma, process

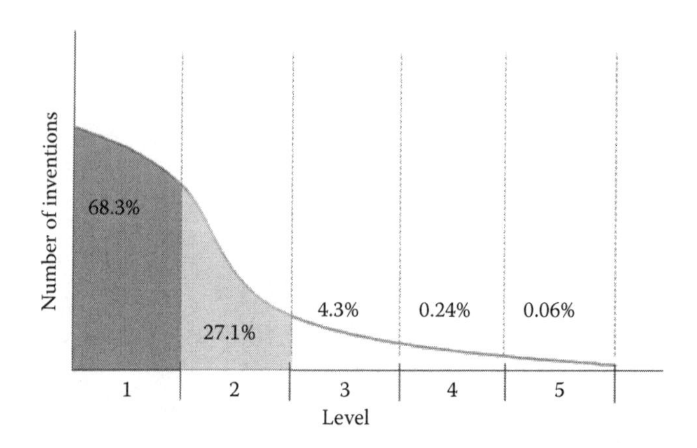

FIGURE 1.3
Percentage of inventions registered in each level of complexity.

reengineering, streamlined process improvement, and Total Quality Management build upon the base where all the people involved have been trained and understand the methodology. Rather than focusing upon education and training, LTM focuses on accomplishing the desired results.

Figure 1.4 shows the five degrees of complexity plotted as a block diagram without LTM applied.

If you add the LTM to the activities that were used to construct the five degrees of design complexity curve, it should have a major impact on the curve. Figure 1.5 is our estimate of how the five degrees of design complexity curve would change as a result of applying the LTM to each of the levels. We believe that a more detailed study needs to be done to get the correct percentages in Figure 1.5, but this shows the concept we are presenting when LTM is focused upon Levels 1 and 2 with some positive impact on Level 3.

When LTM is applied to an obvious and apparent improvement (Level 1), many of the improvements shift up to a minor improvement (Level 2) that is much better than the original improvement. Other Level 1 improvements result in a more robust improvement. Also, some of the minor improvement solutions will move up into the major improvements (Level 3).

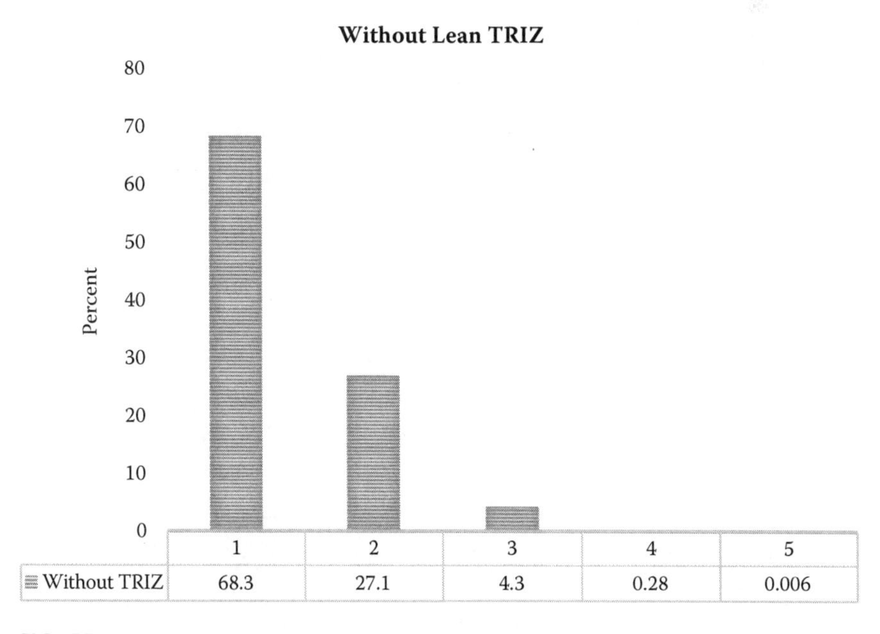

FIGURE 1.4
Five degrees of complexity plotted as a block diagram without LTM.

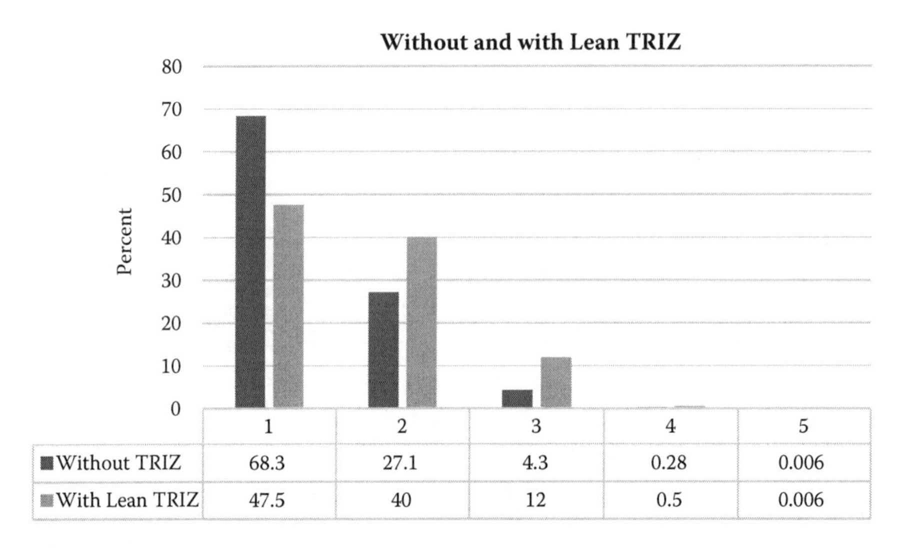

	1	2	3	4	5
Without TRIZ	68.3	27.1	4.3	0.28	0.006
With Lean TRIZ	47.5	40	12	0.5	0.006

FIGURE 1.5
Five degrees of complexity plotted as a block diagram with LTM applied.

We also believe that, if a total classic TRIZ is applied to the appropriate opportunities in Level 3 and Level 4, the percentage of Levels 3 and 4 will be increased significantly. Figure 1.5 shows our estimate of the impact that using LTM would have on each of the five levels. The right-hand bar graph in each of the five quadrants represents the impact that LTM has on the first three quadrants.

Based upon the information presented in Figure 1.5, the following is our estimate of the percent of change from the original five degrees of complexity that can occur when LTM is applied:

- Level 1—Apparent solutions: –30%
- Level 2—Minor improvements: +48%
- Level 3—Major improvements: +179%
- Level 4—New paradigms: +78%
- Level 5—Discovery: 0%

You will note from the preceding list that the quantity of apparent solutions for Level 1 solutions dropped by 30% when LTM was used. This is the desired effect as the decreased Level 1 solutions move up to the more robust solutions in Levels 2, 3, and 4. This often results in the organization coming out with a new product line that is one or two evolutions above the competition.

Typically, the implementation cycle for the product design solutions is 60 days and for the process solutions is 30 days. In this book, we will be talking about a 30-day cycle, but readers should realize that if the opportunity that the LTM team is working on is a design fix or upgrade, it will typically take longer to implement the improvement. Using a 60-day or less window to implement the changes as a result of a design improvement is probably realistic for most organizations.

It should be obvious to all of us that, if over 95% of all of the new ideas fall into Level 1 and 2 categories, focusing on Level 1 and 2 situations to optimize the final solutions can pay huge dividends to the organization. The author agrees that Level 1 and 2 solutions can often be defined using simple brainstorming techniques, but we believe that LTM gives you a much better final solution as it leads the team to consider many alternatives that are normally not considered during a brainstorming session. In about 5% of the opportunities, the organization requires a Level 3 or 4 solution. In these cases, the LTM workshop is extended by an additional two days, allowing the workgroup to use some of the more complex TRIZ approaches.

> To compete in today's fast changing environment, LTM is a very important concept, although not totally new.
>
> **H. James Harrington**

WHY LTM IS DIFFERENT

TRIZ is based upon an extensive knowledge base that was developed by analyzing thousands of patents to determine the thinking pattern that goes into making improvements. All of the five degrees of design complexity presented in the "Five Levels of Invention" section were part of this analysis. (See Figure 1.2.)

The five degrees of design complexity covers the very simplest improvement to the most complex inventions that have been registered at the patent office. Using the TRIZ methodology based upon this massive database stimulates teams to look at individual situations from many different views, allowing them to combine potential solutions together to come up with optimum recommendations. This TRIZ methodology provides the user with a number of proven ways that have been used in the past to solve similar problems and create improved designs. This allows the team

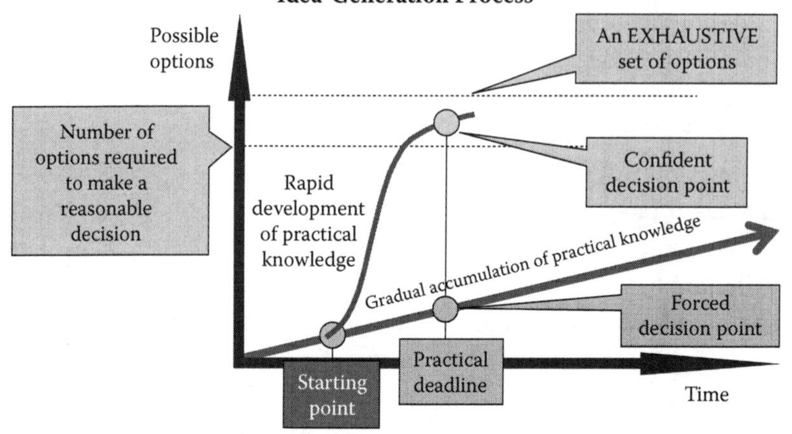

FIGURE 1.6
Comparison of TRIZ approaches compared to the practical knowledge approach.

to come up with superior solutions in far less time than the conventional problem-solving methodologies. (See Figure 1.6.)

SUMMARY

The LTM approach is a very effective way to improve organizational performance and product. The only other approach that comes close to making an equal or better improvement is the information technology approach, which takes a long time to implement and costs a lot of money.

Don't consider LTM as your only improvement methodology. It is a good starting point because it can be completed quickly and has immediate results, but it also leads you into other more advanced TRIZ applications. Look at the results you need to be competitive. If you need a completely new concept, LTM is not designed to do it for you. You will need to use the conventional TRIZ methodology that is designed to attack the more complex situations. We see our clients using the money that they save through the use of LTM to fund other improvement programs.

LTM is the best way to get competitive results fast.

H. James Harrington

2

Five Special LTM Methodologies

Too many people look at a problem and see more work. A wise man looks at the same problem and sees more opportunities.

H. James Harrington

INTRODUCTION

There are five very special tools that we use in LTM. They are the following:

- Value engineering (VE)
- Streamlined process improvement (SPI)
- Lean
- TRIZ
- Lean Process TRIZ (LP-TRIZ)

VE AND VALUE ANALYSIS

Value engineering started during World War II when there were shortages of almost every manufacturing need, from labor to raw materials. While working for General Electric, Lawrence Miles, Jerry Leftow, and Henry Erlicher started the focus on the value of activities. Since 2007, VE users have been encouraged to identify the best practices related to their activity and share them with the other organizations.

Definition: A VE project is not just a collection of studies and best practices; it also includes well-established practices and policies so that they could be integrated into the product development cycle and the improvement projects. VE and value analysis go hand in hand. They are a systematic and organized procedural decision-making process. They are designed to create more value to the organization's stakeholders than was previously provided.

Definition: *Value*—the comparison of function to cost. Value can therefore be increased by improving function or reducing costs.

$$\text{VALUE ANALYSIS} = \frac{\text{Function}}{\text{Cost}}$$

Definition: Value analysis/value engineering (VA/VE) is a systematic and organized decision-making process that evaluates your product and processes from a number of unique perspectives. VE provides its user with a number of potential improvement concepts that the user utilizes to stimulate his or her creativity. It is an effective tool in identifying potential design improvements and minimizing the risk of project failure. It serves as a knowledge base of past best practices within and outside of the organization.

Typical benefits from a VA/VE project are the following:

- Reduce costs—up to 26% across-the-board savings
- Improve operational performance by 40% to 50%
- Improve product quality between 30% and 50%
- Reduce manufacturing costs up to 30%
- Improve customer/supplier relationships
- Major cost avoidance on future projects

(Source: Aether Consulting, Value Analysis/Value Engineering Website (586-939-8028) 8369 Windstone Court, Goodrich Michigan, 48438)

VE concepts are widely accepted as a way to reduce costs, eliminate failures related to engineering design, and reduce the risks related to the development of new products. VE and value analysis are two very different methodologies. VE was developed to aid and alert product engineering, the new product team, and research and development developers about some of the things that should be considered when they are designing new products or improving an old one.

There are many factors that need to be considered in coming up with an optimum design.

You can look at it as a best-practice checklist.

Value analysis, on the other hand, is a detailed study of key processes and products to make them more efficient, effective, and adaptable. It is really a final analysis to estimate the value added of a change.

Typical questions that are asked when conducting VE studies are the following:

- Can the part be eliminated without reducing the functional characteristics of the final assembly and operational reliability?
- Can it be combined with another part or operation?
- Can it be subdivided and included in other operations/parts?
- In the process that produced the part, can it be subdivided into a number of operations?
- Can the operation be conducted during a vital part of another part/component?
- Is the sequence used to produce the item the best possible sequence?
- Would changing the sequence have a detrimental impact?
- Should the activity be performed in another department or on another machine?
- Should an operational flowchart be conducted for the operation to provide a better understanding of its task?
- Is there idle time in the process flow?
- Can the operation be changed or rearranged to make a better match with other parts in the component parts used?
- If the operation is changed, what effects will it have on its interfacing parts and the products' function?
- Will changing the activity change the method used to produce the part or components?
- Can the activity be combined with the inspection of the part or components?
- Is the inspection being performed at the critical part of the process or at the end of the process?
- Will increased inspection reduce cost and improve performance?
- Should the materials used be replaced by less expensive material without decreasing the part's or component's performance?

- Is the quantity of parts provided by the supplier correct to minimize the cost?
- Can the cost related to packaging, moving, transporting, and storing the component's parts be reduced?
- Is the present supplier part of the problem?
- Can the specification be reduced without adversely hurting the final product performance?
- Should the part be produced internally or externally?
- What is the purpose of the parts or components being improved?
- Can part of the parts or components made out of different material be coated to reduce costs or improve performance?
- Are common parts used to reduce confusion in the assembly area?
- Are there better parts or components available on the market?

The examples we listed are very general questions. There are sets of very different questions used for different types of products (e.g., software, printed circuit boards, healthcare).

At present, the VE methodology includes almost 250 different questions. It's absolutely imperative that the individual or team using VE to drive their improvement activities and/or to develop new products has a good understanding of this methodology.

The VE process phases will vary depending upon the application. We have seen as low as 4 and as high as 10. The following is a popular six-phase approach to conducting a VE project:

- Phase I: Data-collection phase
- Phase II: Creative phase
- Phase III: Analysis phase
- Phase IV: Acceptance phase
- Phase V: Implementation phase
- Phase VI: Measurement phase

- *Phase I: Data-collection phase*—The VE/VA starts out by focusing on conducting a function analysis. The function analysis defines what function or performance characteristic is important to the item being improved. Once the functions are identified, data are collected to measure how all these functions are performing prior to improvement.
- *Phase II: Creative phase*—During this phase, the team will generate a number of alternatives that, if accepted, will have the desired impact

on function. Often, a best-practices list of improvement alternatives is used to generate creative ideas by looking at the situation from many different angles.

- *Phase III: Analysis phase*—In this phase, each of the potential improvements will be analyzed to determine how well they meet the required function and the amount of cost savings that would occur if the proposed solution was implemented. This is followed by a value analysis that will actually provide a specific value for each alternative. The teams need to be very careful that the value analysis takes everything into consideration. For example, evaluating one proposed change may reduce the cost of a specific part by 20%, and another proposed change related to the same part will reduce its costs by 40%. This may lead the team to believe that, if both proposed changes are implemented, the cost of the part will be reduced by 60% when, in reality, the combination of the two will only reduce the cost of the part by 40%.
- *Phase IV: Acceptance phase*—During this phase, one or two alternatives are presented to the appropriate management level based upon their responsibilities related to the item being improved. The acceptance of the proposed changes triggers the start of the implementation phase.
- *Phase V: Implementation phase*—This phase starts with the preparation of the necessary paperwork to install the proposed changes. Often, a model is constructed using the official documentation to ensure that the documentation and the proposed change function meet the proposed requirements. A successful completion of the pilot results in the appropriate paperwork being released to the departments that will be implementing the change. This allows the key areas to develop fixtures and tools and obtain parts from the proposed suppliers. The end result of this phase is output that can be sent to or impacts the organization stakeholders.
- *Phase VI: Measurement phase*—During this phase, the functional parameters and cost to produce the output are evaluated to determine if the projected value of the change process is met.

Best practices are a very important part of the VE cycle. We recommend to our clients that they maintain a best-practice listening as part of their knowledge management system. As technologies change and new approaches are developed, these best practices should be added to the organization's best-practice list. They capture past experiences. For example,

- What happens to function if the item was eliminated?
- Does the total item need to be plated?
- Does it need to run that fast?
- Can the same type and size of screws be used throughout the item?
- Can we minimize the number of the different types of screws used?
- Can we use a cast part in place of a machined part?
- Can we use a plastic-molded part in place of a metal part?

Remember that the best-practice lists will vary based upon the industry the list is being used in. For example, the best practices for an automotive industry are very different from the best practices used in healthcare.

The typical places where a list of best practices are maintained include the following:

- American National Standards Institute (ANSI)
- Institute of Electrical and Electronics Engineers
- European Medicines Agency
- Food and Drug Administration

For more information on VE, we suggest reading *Value Engineering: A Guidebook of Best Practices and Tools* (2011) by the US Government of the Defense.

SPI METHODOLOGY

The SPI methodology is an advanced approach to designing and refining processes. It was built upon two methodologies: (1) business process improvement and (2) process redesign. The SPI methodology makes use of a five-phase process called PASIC. (See Figure 2.1.) These letters stand for the following:

- P—Planning
- A—Analyzing
- S—Streamlining
- I—Implementing
- C—Continuous improvement

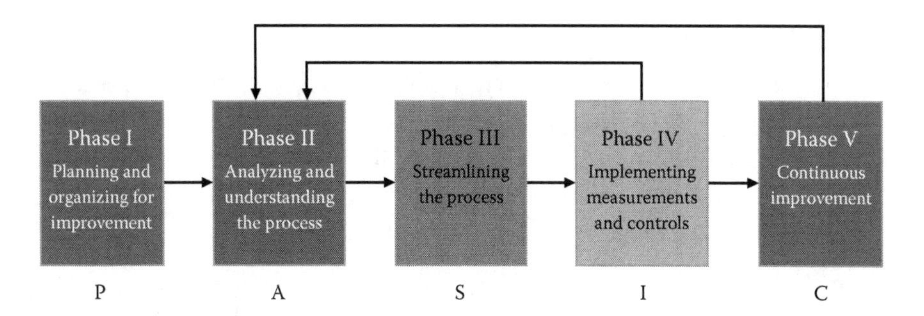

FIGURE 2.1
Five phases of the SPI methodology. (© 2007, Harrington Institute, Inc.)

The following are the characteristics of a well-managed process:

- *Performance*—Produces output on schedule at a lower price than was projected.
- Well-defined boundaries.
- Documented procedures, work tasks, and training requirements.
- Measurement and feedback controls close to the task.
- Customer- and supplier-related measurements and targets.
- Known cycle times.
- Formalized change procedures.
- Copied by other organizations.
- Always looking for ways to improve.

SPI Phase I—Planning

During Phase I of the SPI improvement, the team focuses on reviewing all the major processes and, in concert with the executive team, defines the processes that should be streamlined. These processes are then put in priority order, and teams are assigned to work on the one to three of highest-priority processes. Separate teams are formed for each of the priority processes made up of individuals involved in the process to be studied. The team then defines the boundaries of the process and establishes efficiency, effectiveness, and adaptability measurements for the selected processes. In addition, a detailed project plan is prepared and approved by the management team. Many of these activities will have already been completed at this point in the LTM.

SPI Phase II—Analyzing

For process improvement opportunities, the LTM team will primarily focus on SPI Phase II (analyzing and understanding the process) and Phase III (streamlining the process). During Phase II, the primary tools that will be used are process walk-through and flowcharting. There are many different types of flowcharts. Figure 2.2 lists seven of the ones that are most frequently used.

Figure 2.3 is an example of a functional flowchart that is often called a swim lane flowchart. In order to provide a view of a more complex process, a flowchart can be developed. In these cases, information about each of the boxes in the flowchart is recorded on the right-hand side of the paper. Typical things that are recorded are the name of the activity, relative cycle time and processing time, defect rates, and costs.

Figure 2.3 is a typical swim lane flowchart that includes processing time and cycle time. Figure 2.4 is a typical graphic flowchart. These charts are used to show movement between different locations. The objective of

	Level of detail	What is charted	What it includes
1. Process blocks	High	Activities	What
2. Process charts	Medium	Activities Tasks Decisions	What, how, where
3. Procedure charts	High	Tasks Material flow Individual actions	How, where, who
4. Functional flowcharts	Medium	Activities Tasks	What, who
5. Geographic flowcharts	Medium	Activities	What, where
6. Paperwork flowcharts	Medium	Documents	How, where, who
7. Informational flowcharts	Medium	Information Activities Decisions	What, how, who, where

FIGURE 2.2
Seven different types of flowcharts.

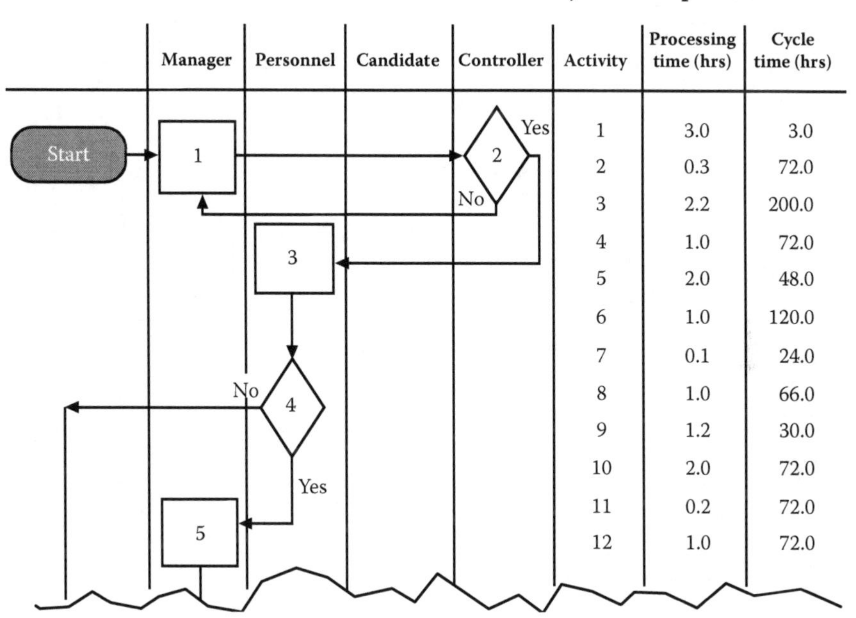

Activity	Processing time (hrs)	Cycle time (hrs)
1	3.0	3.0
2	0.3	72.0
3	2.2	200.0
4	1.0	72.0
5	2.0	48.0
6	1.0	120.0
7	0.1	24.0
8	1.0	66.0
9	1.2	30.0
10	2.0	72.0
11	0.2	72.0
12	1.0	72.0

FIGURE 2.3
Functional flowchart/swim lane flowchart.

these flowcharts is to look for ways that can minimize transportation cost and cycle time. Activities like one unit built, one-minute changeover, the relocation of serial activities, and combining activities make one person responsible for more activities.

Graphic flowcharts are used to track materials or people flow during the process. Minimizing the movement of parts and/or people can have a big impact on both costs and cycle time. Normally, organizations worry about processing time as it is very costly. Customers are not concerned with processing time but are highly impacted by cycle time.

There are many different kinds of flowcharts that can be used depending upon the individual process being evaluated. Some of the more popular ones are as follows:

- ANSI standard flowcharts
- Simple business process flowcharts
- Functional flowcharts
- Functional timeline flowcharts
- Graphic flowcharts

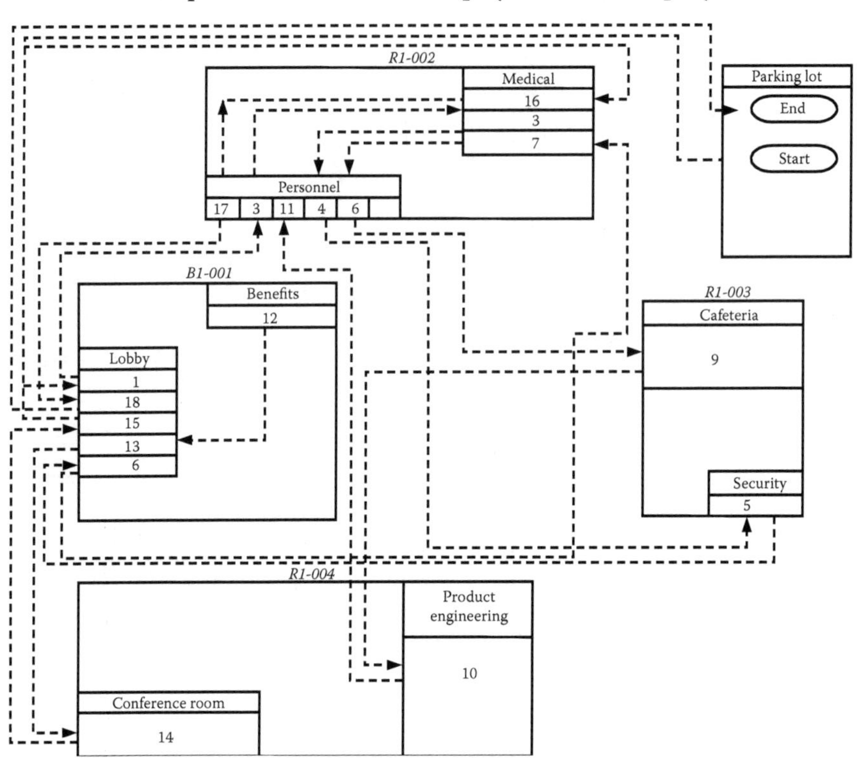

FIGURE 2.4
Typical graphic flowchart.

- Knowledge management flowcharts
- Communications flowcharts
- Value stream mapping

The flow of work between departments becomes a costly and time-consuming activity. Figure 2.5 is a typical current-state value stream diagram (often called a value stream map). Figure 2.6 is a future-state value stream diagram of the same process. We commonly use a data swim lane flowchart (function data flowchart) in place of the value stream diagram because it is quicker to prepare and it contains a lot more information. (See Figure 2.7.) When this is the case, a product flowchart between departments is an excellent way to highlight this type of problem. Figure 2.8 is a typical process sequence chart. Figure 2.9 is a process sequence chart filled out with all its information. You'll note that it lists

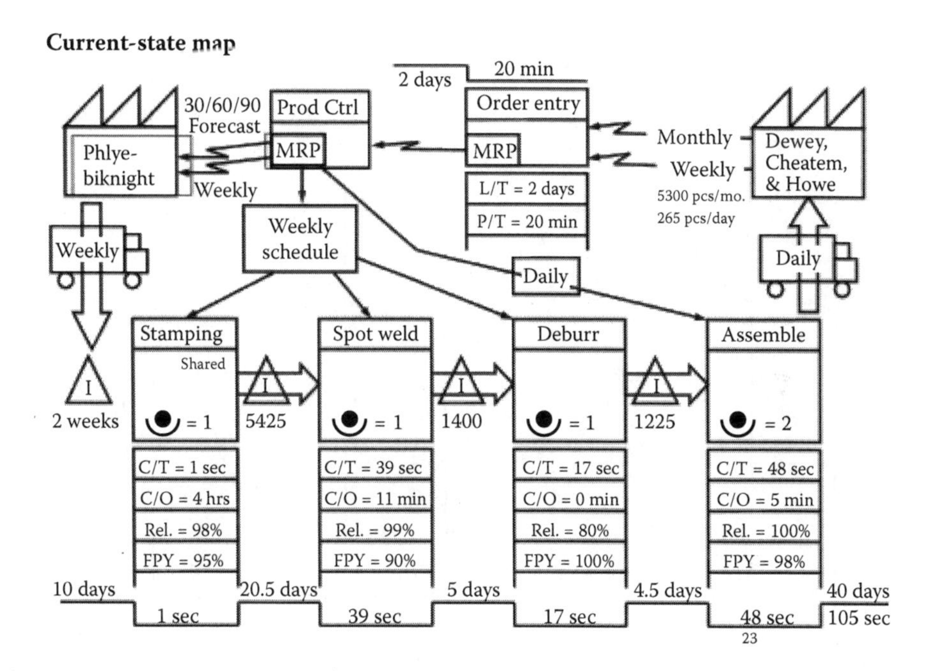

FIGURE 2.5
Typical current-state value stream diagram.

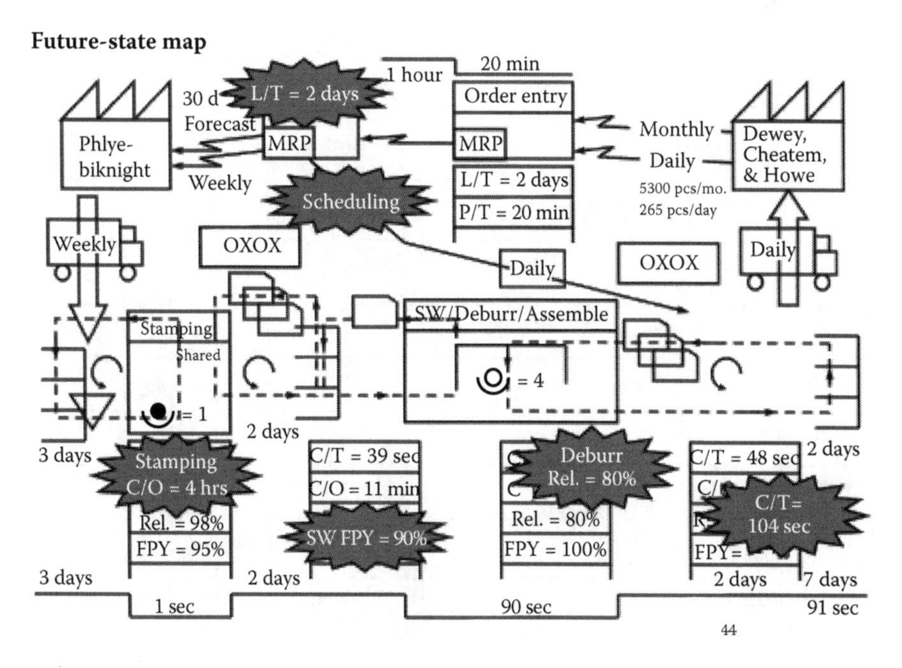

FIGURE 2.6
Example of a future-state value stream diagram.

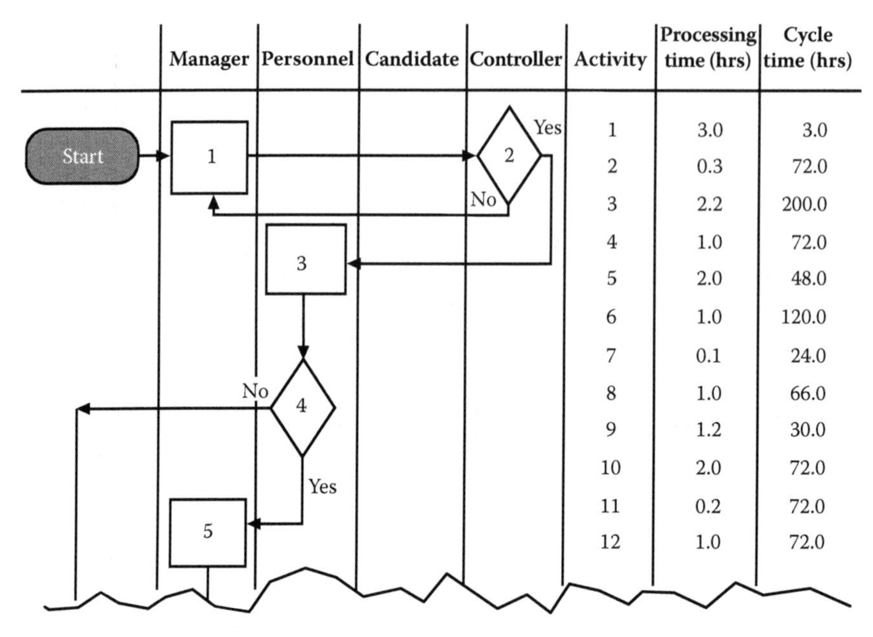

FIGURE 2.7
Function data flowchart.

Process sequence chart

Movement of forms (245 and 355)	**Current performance duration (mins)**					
Inspection forms 245 and 355 and client's file (collected from files)	5–15 mins					
245 and 355 placed on trolley basket (batches) Delay awaiting transfer	} 10–20 mins					
Trolleyed to section leader's desk (batches)	0.1 min					
Wait for signature	5–10 mins					
245 and 355 inspected and signed	1 min					
Await transfer to final preparation	10–15 mins					
245 and 355 taken to final preparation desk (batches)	0.5 min					
Await final preparation	10–12 mins					

FIGURE 2.8
Process sequence flowchart.

Map the Activity

PROCESS FLOW CHART

Sheet No1 of1

Department Orders Rec. Chart By TSF

Process Name Processing an order Date 7/02/99

____ Proposed Method ____ Present Method

No.	Chart Symbols	Cycle Time	Process Time	Process Description
1		12 H	4 M	Receive order by phone, fax, or mail
2		1 H	5 M	Write phone order onto form
3		1 H	3 M	Check fax and mail orders for errors
4		1 H	1 M	Is the order correct?
5		12 H	6 M	Contact customer to get required data
6		14 H	4 M	Price order
7		5 H	3 M	Make copy of order and price sheet
8		16 H	2 M	File original
9		30 H	3 M	Check credit limits with bank
10		1 H	4 M	If credit check is good, enter order in computer
11		.1 H	2	Send order copy to warehouse
12		1.5 H	0	Wait for return of order copy from warehouse. Hold until warehouse returns order copy.
13		1 H	1 M	Check to see if the order could be filled
14		.1 H	3 M	Modify order to show back order and provide customer with a target delivery date
15		.1 H	2 M	Print out shipping documents
16		24 H	2 M	Send shipping documents to warehouse
17		.1 H	2 M	Print out invoice

O = Start or stop the process ◯ = Inspect ⇨ = Movement ▭ = Operation ◇ = Decision point ⊐ = Delay ▽ = Storage

FIGURE 2.9
Total process sequence flowchart.

all of the 17 activities in the process with the cycle time and processing time for each. The preprinted form allows the user to draw lines between activities showing what kind of activity it is. Figure 2.10 shows the product flow from department to department. Often, the same type of chart is used to show product flow from machine to machine and/or operator to

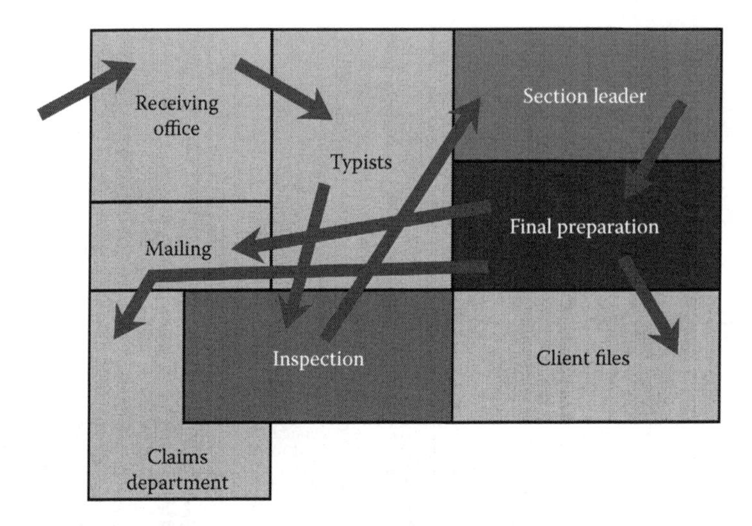

FIGURE 2.10
Product flow from department to department.

operator. Figure 2.11 is the new product flow for the same product after it has been streamlined.

Using flowcharts will also help you to identify the key part of the process that limits the process output capabilities. These throughput limiting operations are excellent points to start any improvement action.

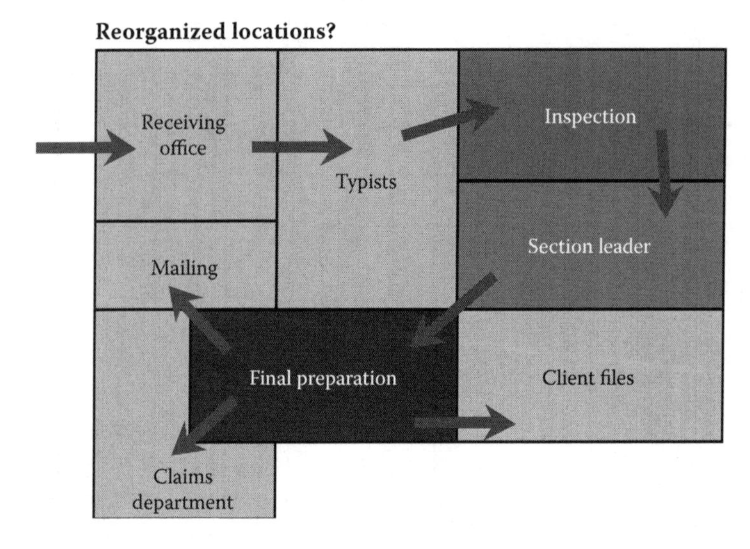

FIGURE 2.11
Product flow from department to department after a product flow has been reorganized.

SPI Phase III—Streamlining the Process

Phase III (streamlining the process) is the most important activity from the LTM methodology that is directed at process improvement. During this phase, the process that is being studied is subjected to 12 different tools with the objective of maximizing its performance. The 12 tools are as follows:

1. Bureaucracy elimination
2. Value-added analysis (VAA)
3. Duplication elimination
4. Simplification
5. Cycle time analysis
6. Error preventing
7. Supplier partnerships
8. Technology
9. Process upgrading
10. Risk management
11. Standardization
12. Simple language (improve communication)

Apply the Streamlining Approaches

This approach takes the present process and removes waste while reducing cycle time and improving the process effectiveness. After the process flow is streamlined, automation and information technology are applied, maximizing the process' ability to improve its effectiveness, efficiency, and adaptability. SPI is sometimes called *focused improvement* or *process redesign* since it focuses on an individual process. (See Figure 2.12.)

- *Bureaucracy elimination*—The streamlining activities will typically start by focusing on bureaucracy elimination. We like to start with bureaucracy elimination because everyone wants to get rid of the bureaucracy within the organization. Management and employees lightly jump on the bandwagon of eliminating or at least minimizing the bureaucracy that is built into the organization's processes. We start this activity by looking for duplicate activities and/or additional no-value-added (NVA) reviews and/or approvals. We then do an analysis to determine if the bureaucracy can be justified as part of the process when you consider cost and cycle time additions. Most of

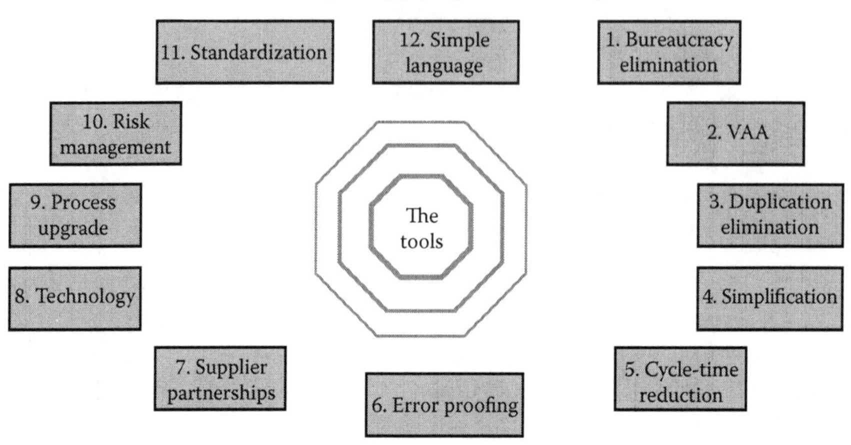

FIGURE 2.12
The 12 activities in streamlining.

the bureaucracy activities, when subjected to a critical review, cannot be justified. All too often, the checks and balances in a process are directed at identifying 0.5% of the people in the company who are dishonest. The other 99.5% are punished because management is reluctant to step up and get rid of the dishonest individuals.

- *VAA*—The LTM team will start off by focusing on the process flow diagram with the objective of removing unnecessary NVA activities, minimizing business-value-added activities, and maximizing real-value-added (RVA) activities. This is called streamlining the process (SPI). VAA is very different from VE. VE methodology is basically a series of questions used by product engineering to identify weaknesses in a product design. VAA is an analysis of every activity on the process flowchart to determine its contribution to meeting end-customer expectations. The objective of VAA is to optimize RVA activities and minimize or eliminate NVA activities. The organization should ensure that every activity within the business process contributes real value to the entire process. Value is defined from the point of view of the external customer's standpoint. (See Figure 2.13.) There are three classifications of value activities. They are as follows:
  1. *Real-value-added activities*—These are the activities that, when viewed by the external customer, are required to provide the output that the customer is expecting. There are many activities

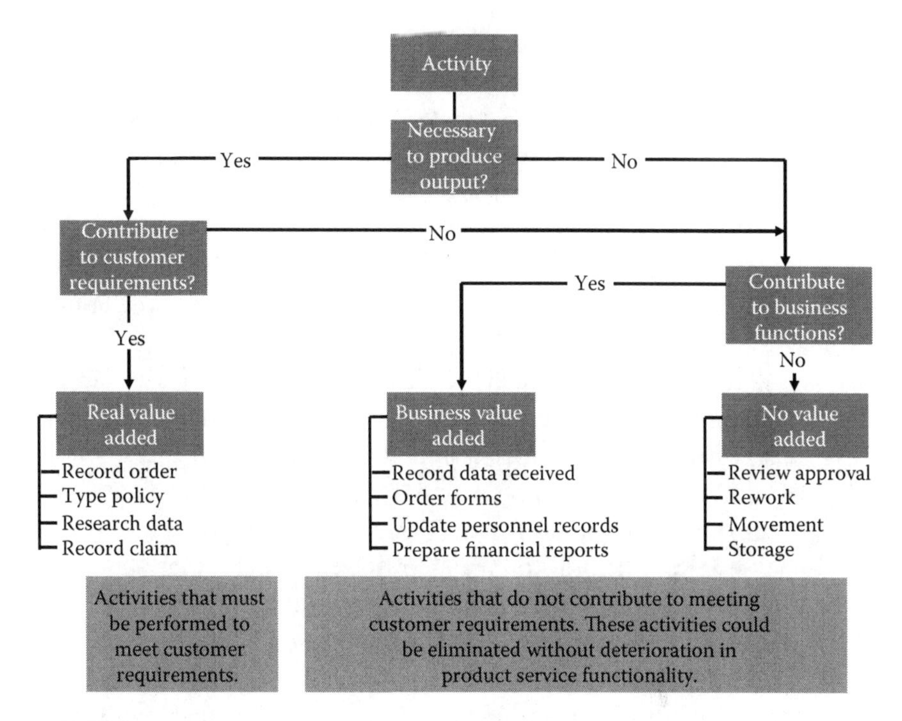

FIGURE 2.13
Value-added assessment.

performed that are required by the business but that are NVA from the external customer's standpoint.

2. *Business-value-added activities*—These are activities that need to be performed in order to run the organization but that add no value from the external customer's standpoint (e.g., preparing budgets, filling out employee records, updating operating procedures).
3. *No-value-added activities*—These are activities that do not contribute to meeting external customer requirements and could be eliminated without degrading the product or service function of the business (e.g., inspecting parts, checking the accuracy of reports, reworking a unit, rewriting a report). This includes activities classified as bureaucracy activities. There are two kinds of NVA activities:
   a. *Activities that exist because the process is inadequately designed or the process is not functioning as designed.* This includes movement, waiting, setting up for an activity, storing, and doing work over. These activities would be unnecessary to produce the output of the process but occur because

of poor process design. Such activities are often referred to as part of poor-quality cost.

b. *Activities not required by the external customer or the process and activities that could be eliminated without affecting the output to the external customer, such as logging in a document.*

Every organization has a huge hidden office made up of BVA and NVA activities. It often accounts for 80% of the total effort, while RVA activities only account for 20% of the organization's total effort. (See Figure 2.14.)

RVA activities contribute directly to producing the output required by the external customer. The LTM team should analyze each activity and/or task on the flowchart and classify it as an RVA, a BVA, or an NVA activity. (Note: the bureaucracy activities will also be classified as BVA or NVA activities.)

It is very helpful to create a rainbow flowchart, as shown in Figure 2.15. Since this book is printed in black and white, you can't actually visualize the benefits of seeing different colors. However, in the actual workshop, it would be extremely beneficial to create this type of flowchart. To start, use a yellow highlighter to designate each BVA activity on the flowchart. Then, color in the NVA activities with a red highlighter, and, next, color the bureaucracy type activities in blue. You have now turned your flowchart into a rainbow flowchart. Typically, as LTM team members go through this phase of the analysis, they are astonished at

FIGURE 2.14
Picture of hidden office. (© 2007, Harrington Institute, Inc.)

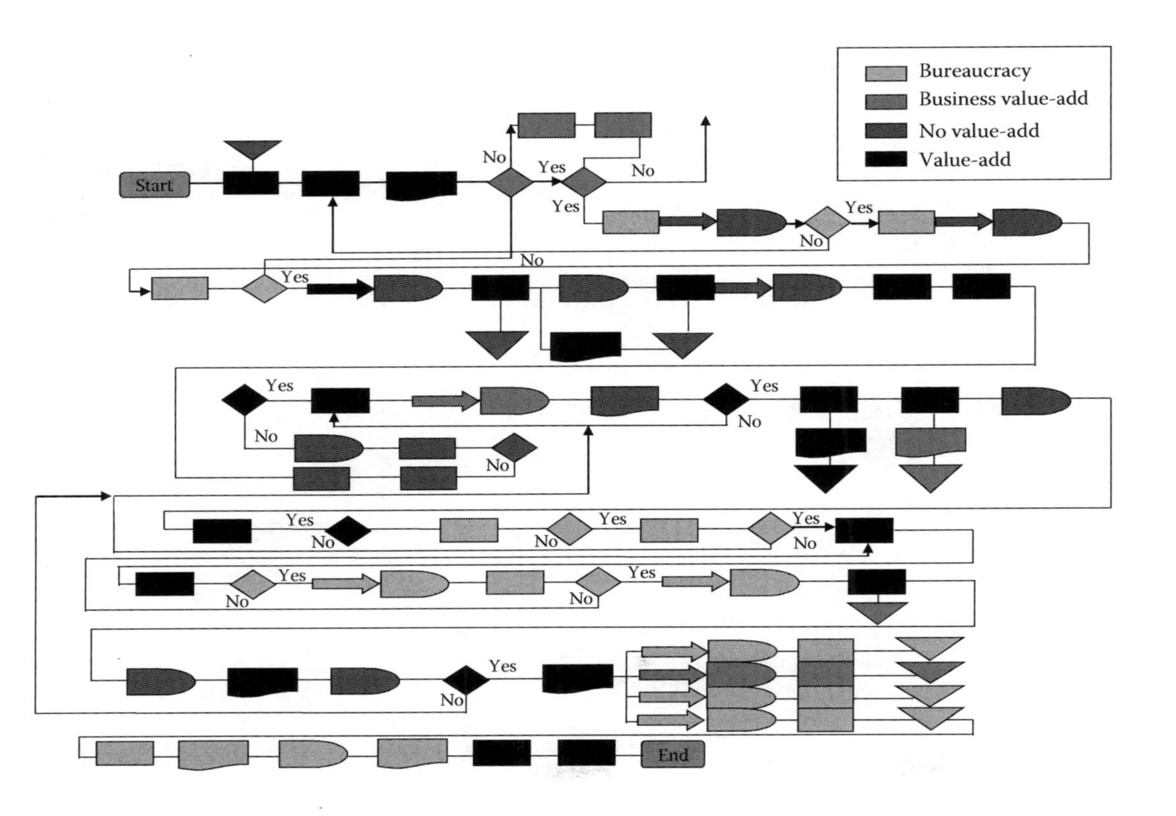

FIGURE 2.15
Rainbow flowchart.

the small percentage of costs that are RVA activities. Even more alarming is the mismatch of processing time for RVA activities compared to total processing time. For most business processes, less than 30% of time is spent in RVA activities.

Obviously, this indicates something very wrong, and managers are often disturbed when they learn of these numbers. But there are several explanations:

- As the organization grows, processes break down and are patched for use, thereby making them complex.
- When errors take place, additional controls are put in place to review outputs rather than change the process. Even when the process is corrected, the controls often remain.
- Individuals in the process seldom talk to their customers and hence do not clearly understand the customer's requirements.
- Too much time is spent on internal maintenance activities (such as coordinating, expediting, and record keeping) instead of doing RVA work.

• *Duplication elimination*—With this tool, we eliminate any duplication efforts going on within the process. This includes databases and information put into the database.
• *Simplification*—With this tool, we look for ways where we can simplify documentation and activity flow. This reduces processing time and costs while improving the quality of the output.
• *Cycle-time reduction*—In this activity, management focuses its efforts and attention on reducing processing time because processing time is very costly. But, from a customer's standpoint, the only time-related measurement is cycle time. An organization's reputation is based upon output quality and cycle time.
• *Error proofing*—In this activity, the group brainstorms how they could cause the process/product to fail and then develops improvement activities that will eliminate or minimize the failure on the process and organization as a whole.
• *Supplier partnerships*—This activity involves establishing close working relationships with all your primary suppliers. Best results occur when there is a close positive working relationship between the customer and the supplier.
• *Technology*—Technologies are advancing so rapidly that it is easy to lose market share because technology has changed. This includes software and automation.

- *Process upgrade*—Little things are important to your employees like the following: what color the walls are painted, the layout of the office, the filing system that is used, the clean desk policy, the room they have to work on, how fast their computer works, the location of their desk, how comfortable their chairs are, etc. All of these are important to the workers, and they impact their productivity. Upgrading the work environment is often the key to improving process flow and reducing process costs.
- *Risk management*—This is a tool that is used to evaluate and minimize the risks related to improvements that are scheduled to go into the process. By minimizing risks, you've greatly increased the probability of having a successful project.
- *Standardization*—If you can standardize the activities, then they can be improved. When everyone is doing it a different way, it's extremely hard to bring about improvement. When you standardize the one best way, it is possible to make real improvement because the problem is then with the process rather than the individual.
- *Simple language*—All too often, we write to impress our counterparts in the documentation. Improving communications is one of the best ways to improve individual performance.

  Most business writing at present cannot be read or understood easily. After a cursory glance, these writings are usually routed to the nearest file cabinet or trash can. While the current world makes receiving accurate and timely information more important than ever, the quality of most of the business writing lingers in the dark ages. It is pompous, wordy, indirect, vague, and complex. One critique noted, "Too often, business reports are wanting everything but size."

  The LTM team needs to evaluate the present documentation used in the process to ensure that it is written for the user. In the LTM team's effort to simplify communication, they look at things like the following:

  - *Need to determine the reading comprehension level of the audience.* The documentation should be written so that all readers can easily understand the message. If the reader has an 8th grade comprehension level, prepare the documents for the 7th grade. And just because your audience all graduated from high school doesn't mean that they can read and comprehend at the 12th grade level. Many college graduates' reading comprehension level is below the

10th grade. When English is a second language, the reader's comprehension level is often much lower than the general education level. When writing for people whose second language is English, write at a level below their general education level, and always use the dictionary's first preference meaning only.

- *Need to determine how familiar the audience is with the terms and abbreviations that are used in the document.* Unless it is critical to the work assignment, don't use the new terms or jargon. If it is necessary to use these words, be sure to clarify and define them.
- *All procedures that are more than four pages long should start with a flowchart to help the reader through the details within the procedure.*

During the 12 streamlining activities, the LTM team needs to keep asking itself the following questions:

- How can the RVA activities be optimized?
- Can the RVA activities be done at a lower cost with a shorter cycle time?
- How can the NVA activities be eliminated? If they cannot, can they be minimized?
- Do we have to spend too much time on BVA activities? If so, how can we minimize it?

LEAN

It's difficult to talk about Lean. Initially, it was very simple as it focused on picking up, cleaning up, and putting away, thereby eliminating clutter along with the use of some simple improvement tools. Over the past five years, Lean keeps expanding. There's hardly a day that goes by where some ambitious consultant or engineer doesn't add one of the Total Quality Management (TQM) tools to the Lean body of knowledge. As a result, the Lean program continues to change and grow in complexity.

The Lean approach is based upon Toyota's manufacturing process. Although most of the tools are adaptations from ones that were already in use, the combination that Toyota put together allowed Toyota to produce an outstanding car in an extremely short cycle time at low cost and big profits.

The Lean methodology, as it developed, is based upon the following five concepts:

1. Lean thinking
2. Understand what your customers value
3. Create a culture of continuous improvement
4. Use data to drive improvement
5. Identify and eliminate waste

The Lean methodology started with the concept known as the Five S's. It focused on keeping a neat work area and putting things away where they belong. It also focused on getting rid of items that were not used on a regular basis. Currently, it has widened to mean the elimination of all waste.

The elimination of waste cycle usually starts by the LTM team conducting a brainstorming session directed at identifying waste from this operation and equipment. To do this, the facilitator will lead the team through a brainstorming session where different colored Post-its or tabs (cards) are used. Of course, this is just one of the many derivations that are used in conducting a brainstorming session. All of these deviations will produce acceptable results in this case. As a result, the facilitator should select one that he or she is most comfortable with. We personally like to use the colored tag approach, which we will be discussing here. The facilitator will start with the green dot cards and lead the team through a brainstorming session to define the improvement actions. These improvement actions are recorded on index cards. The same approach is then applied to the yellow- and orange-colored index cards until the allotted time is used up. This is a very time-consuming task, and it will require everyone to be creative and original thinkers. The facilitator will encourage the team to define improvement actions that are easy to implement by the team members. Ideas that are outside of the scope of the LTM project should be recorded as they provide good input to the management team. The biggest problem the facilitator faces during this activity is to keep the team saying, "*This is what we can do!*" rather than, "*This is what they should do.*" Be sure not to rule out a good idea just because it had failed before. Sometimes, conditions have changed, and things that did not work before will work now.

The facilitator should point out to the team that they should look for time wasters like reports, approvals, meetings, politics, practices, etc. Then, think about each of the items to see if they could be eliminated, partially eliminated, delegated downward, done less often, done in a less complicated way or time consuming manner, or done with fewer people. Often, although an activity may have RVA, big parts of it are really NVA. Each card should have the name of the person that filled out the card on it. The cards will be placed under the root cause or contradiction they relate to. In most cases, there are two slightly different approaches used—one approach for defining improvement activities for processes and a slightly different approach for defining improvement activities for evolutionary designs.

The following are some of the Lean tools that may be considered for use in improving the processes in your organization:

- *Andon*

  Andon is a system that provides visual feedback to management and employees relating to the status of production. It is also used as a way to notify other departments that help is needed and is often used to allow employees to stop the production line.
- *Automation*

  Automation is the point in the process that uses software, equipment, or robotics that complete an activity that is normally done by an operator. It has the advantage of its output being more *consistent than in the same operation being done by a human being.*
- *Bottleneck analysis*

  This is the point in a process flow that limits the number of units that can be manufactured. It is the activity/operation that limits the maximum number of units that can be produced in a continuous flow process activity.
- *Continuous flow*

  Continuous flow is a description of a manufacturing process with a minimum amount of stored material throughout the process. Ideally, when one operation is completed, the next operation is started immediately without the unit being stored.
- *Error proofing*

  Error proofing is the process of designing products and processes so that it is difficult for the operator and automation to produce an output that is not to specifications. Typically, it is a pin in a fixture that does not allow the operator to insert the part in the wrong direction. In the design area, it is designing parts so that they can only go together one way.

- *Just-in-time*

  Just-in-time (JIT) describes a condition where storage is eliminated. It exists in processes where there are no delays between activities. The suppliers supply parts as they are needed on the production line. In practice, it is normal not to have more four hours backup for any activity. JIT is a myth in most organizations. If Toyota's system was a JIT system, they would be assembling each car for a specific customer. In truth, they are building to a ship schedule. Probably, Dell Computer is a better example of JIT. At least, their final assembly is designed to meet a specific customer's needs. We don't consider JIT if the suppliers have to have a buffer in case there is a demand for additional units from their customer.
- *Overall equipment effectiveness*

  Overall equipment effectiveness is the measurement system that is primarily a measure of the manufacturing process, but, in the current environment, efficiency in the support and white-collar areas becomes a major objective as customers now view the product cycle, starting when the order was accepted until the item is delivered and working properly. The measurement system that is used in production to measure the effectiveness of a process or activity is typically made up of measurements related to availability, quality, and performance.
- *Pull system*

  A pull system is one that produces parts to a specific customer's order. They work to specific customer requirements rather than a production schedule. For example, in the 077 process at IBM, each machine had the name and address attached to it when the frame started through the assembly and test process. More and more, currently, customization is becoming the best route to success.
- *Six big losses*

  Six big losses indicates the six ways that production systems most often have problems with. They are the following:

  1. Breakdowns
  2. Setup/adjustments
  3. Small stops
  4. Reduced speed
  5. Startup rejects
  6. Production rejects

In a performance improvement effort, action needs to be taken to minimize the impact that each of the big six losses has on the organization.

- *Takt time*

  Takt time is the rate at which a finished product needs to be completed in order to meet customer demand. If a company has a takt time of five minutes, it means every five minutes, a complete product, assembly, or machine is produced off the line because on average a customer is buying a finished product every five minutes. If the sales rate was two per hour, then the ideal takt rate would be two per hour. The objective is to have a build rate that is in harmony with the sales rate. The ideal situation would be when every manufacturing operation's throughput rate was equal to the sales rate. This does not mean that the organization needs to have a JIT system. Typically, at the end of the process, a buffer stock is stored, as well as the buffer stock for supplier-provided parts and equipment.
- *Visual factory*

  The visual factory is primarily a manufacturing approach to provide the operator with key information that is related to a specific product. It could be a visual screen that shows how many units have completed the production process for a specific time, along with the target production rate for that same period. For example, in the service industry, it can show the number of calls waiting in a service department and/or the average length of time that is being taken to discuss the customer problem with the customer.

Some additional tools to consider include tools and methodologies used in the Six Sigma programs. Of course, the Six Sigma program is less robust than the TQM program, which was popular back in the 1980s.

- A3 problem-solving
- Efficient movement
- Ishikawa diagrams
- First in first out
- Key performance indicators
- Process engineering
- Lean Six Sigma
- Lean software development
- Lean thinking
- Continuous improvement

- Spaghetti plots
- Total productive maintenance
- Value stream mapping

TRIZ

In order to gain an understanding of TRIZ, the reader must first gain an understanding of classical TRIZ and its background. Classical TRIZ is the brainchild of a brilliant Russian scientist, Genrich Altshuller. Altshuller was born in the former Soviet Union in 1926. At the age of 14, he created his first invention for improving equipment for scuba diving. His job in the Soviet navy was to help inventors apply their patents. His involvement in problem-solving related to new concepts that stimulated him to find new techniques for this process. In an effort to improve his problem-solving abilities, he started analyzing patents in order to find how inventors created inventions. As he became a member of the Russian patent office, he had access to the details related to patents that were issued around the world. He reasoned that, through analyzing patents, he could find answers to how thousands of developers and scientists solve complicated problems and create innovative products. The study of inventions became the starting point for the creation of the theory of invention problem-solving (TRIZ), which is a Russian acronym for the theory of inventive problem-solving. (See Figure 2.16.)

In an effort to define the thought pattern that led to creative invention, Altshuller reviewed over 200,000 patents with the objective of identifying the thought patterns that supported their creation. This analysis reduced the 200,000 patents down to 40,000 that had inventive solutions. The rest were minor improvements. The basis analysis was on a set of conditions he called *system contradictions.* He defined system contradictions as an inventive problem in which the new parameter that changes was in conflict with another parameter of the process or product. His studies led him to define 39 frequently used parameters that could cause system contradictions in problems, products, and processes. The 39 characteristics is a list of 39 engineering parameters for expressing technical constraints defined in the late 1960s. (See Figure 2.17.)

For example, in Figure 2.17, number 9 is *speed.* It indicates that you can speed something up or slow down to offset some of the other technical constraints. Number 23 is *waste of substance.* In this case, you need to

What is TRIZ?

- Systematic, structured way of thinking, analyzing, and innovating
- A repeatable methodology
- Extensive knowledge base

Genrich Altshuller
10/15/26–9/24/98

FIGURE 2.16
TRIZ overview.

39 engineering parameters for expressing technical contradictions

1 *Weight of moving object*
2 *Weight of nonmoving object*
3 *Length of moving object*
4 *Length of nonmoving object*
5 *Area of moving object*
6 *Area of nonmoving object*
7 *Volume of moving object*
8 *Volume of nonmoving object*
9 *Speed*
10 *Force*
11 *Tension, pressure*
12 *Shape*
13 *Stability of object*
14 *Strength*
15 *Durability of moving object*
16 *Durability of nonmoving object*
17 *Temperature*
18 *Brightness*
19 *Energy spent by moving object*
20 *Energy spent by nonmoving object*
21 *Power*
22 *Waste of energy*
23 *Waste of substance*
24 *Loss of information*
25 *Waste of time*
26 *Amount of substance*
27 *Reliability*
28 *Accuracy of measurement*
29 *Accuracy of manufacturing*
30 *Harmful factors acting on object*
31 *Harmful side effects*
32 *Manufacturability*
33 *Convenience of use*
34 *Repairability*
35 *Adaptability*
36 *Complexity of device*
37 *Complexity of control*
38 *Level of automation*
39 *Productivity*

FIGURE 2.17
The 39 parameters of technical contradictions. (© 2000–2008 Idealon Internationl Inc., All Rights Reserved.)

analyze the product to determine which substance is really not necessary or the amount of the material that can be reduced.

Altshuller was then faced with the problem of when a system contradiction occurs, what can be done to offset the negative impact? This led him to define 40 fundamental inventive principles that were used over and over again to offset the negative impact related to the system constraints. (See Figures 2.18 and 2.19.) The definition of the 40 TRIZ principles is that they are statements that describe approaches to resolving technical conflicts (problems and/or contradictions) that were defined by Altshuller based upon his study of over 200,000 patents. These 40 TRIZ principles have a twofold purpose:

1. Within each principle resides guidance on how to conceptually or actually change a specific situation or system in order to get rid of a conflict.
2. The 40 principles also train users in analogical thinking, which is to see the principal as a set of patterns of inventions or operators applicable to all fields of study. (See Figure 2.20.)

40 inventive principles (1–20)

Principle 1 Segmentation
Principle 2 Taking out/extraction
Principle 3 Local quality
Principle 4 Asymmetry
Principle 5 Merging/consolidation
Principle 6 Universality
Principle 7 *Nested doll*
Principle 8 Antiweight
Principle 9 Preliminary antiaction
Principle 10 Preliminary action
Principle 11 Beforehand cushioning
Principle 12 Equipotentiality
Principle 13 Do it in reverse/*the other way around*
Principle 14 Spheroidality/curvature increase
Principle 15 Dynamics
Principle 16 Partial/excessive actions
Principle 17 Transition to another dimension
Principle 18 Mechanical vibration
Principle 19 Periodic action
Principle 20 Continuity of useful action

FIGURE 2.18
40 TRIZ principles. (TRIZ Solutions LLC copyright © all rights reserved.)

40 inventive principles (21–40)

Principle 21 Rushing through/skipping/hurrying
Principle 22 Converting harm into benefit
Principle 23 Feedback
Principle 24 Intermediary/mediator
Principle 25 Self-service
Principle 26 Coping
Principle 27 Cheap, short-living objects
Principle 28 Mechanical interaction substitution
Principle 29 Pneumatics/hydraulics
Principle 30 Flexible shells and thin films
Principle 31 Porous materials
Principle 32 Color changes
Principle 33 Homogeneity
Principle 34 Rejecting and regenerating parts/discarding and recovering
Principle 35 Parameter changes
Principle 36 Phase transition
Principle 37 Thermal expansion
Principle 38 Accelerated oxidation/strong oxidants
Principle 39 Inert atmosphere
Principle 40 Composite materials

FIGURE 2.19
40 TRIZ principles (continued). (TRIZ Solutions LLC copyright © all rights reserved.)

It should be obvious to you that it would be extremely difficult to work with the many combinations that are created between the 40 principles and the 39 engineering parameters for expressing technical constraints. There are actually hundreds of potential combinations between these two problem-solving approaches. In order to simplify and to guide the people using the TRIZ methodology, Altshuller developed a matrix of engineering parameters for expressing technical constraints and the 40 TRIZ principles. This matrix consisted of listing the 39 engineering parameters for expressing technical constraints along the vertical and horizontal lines. The vertical line is used to define the engineering parameters for expressing technical constraints that you are considering using (Improving Parameters). Across the vertical part of the matrix, the same 39 engineering parameters for expressing technical constraints is listed, but, this time, they are the 39 constraints that could have a negative impact upon the outcome (worsening parameters).

Definition: Matrix of system constraints—It makes use of the 39 engineering parameters for expressing technical constraints and 40 TRIZ principles. Both the vertical and horizontal of the matrix is made up of the 39 engineering parameters for expressing technical constraints.

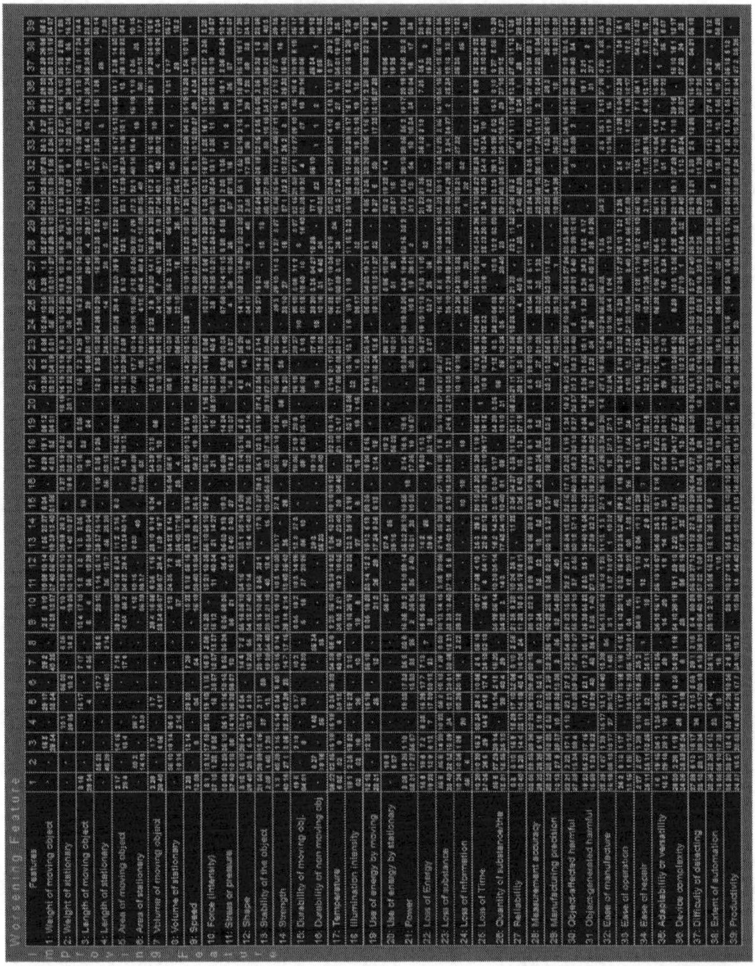

FIGURE 2.20
Matrix of system constraints.

(See Figure 2.20.) Vertical terms are used to pick an action that you want to consider using to improve the item you are working on. The horizontal axis is used to identify negative things that could happen based upon the proposed solution the team is looking at. At the junction of the horizontal and vertical columns, up to four of the TRIZ principles are listed that are the most likely to offset the contradiction based upon past experience.

Figure 2.13 is impossible to read because of the large amount of information that it contains. It is typically printed out on a sheet of paper that is 20 × 20 or on a larger paper, making it very readable. It was presented here just to provide the readers with an idea of how complex it is. To help you better understand the matrix of system constraints, we have taken the first five and made them big enough so you can read them. (See Figure 2.21.)

Don't let this matrix overwhelm you. It's really quite easy to use once you understand the matrix. To help you understand how to use it, we have taken a small part of it and enlarged it. (See Figure 2.21.)

For this example, we will follow an individual engineering parameter for expressing technical constraints—speed. (See Improving Parameters point #1 in Figure 2.22.) When we select speed on the vertical axes as a way to improve the item, you will note that there are 30 of the 39 intersecting

		Weight of a mobile object	**Weight of a stationary object**	**Length of a mobile object**	**Length of a stationary object**	**Area of a mobile object**
		1	2	3	4	5
Weight of a mobile object	1			15, 8, 29, 34		29, 17, 38, 34
Weight of a stationary object	2				10, 1, 29, 35	
Length of a mobile object	3	8, 15, 29, 34				15, 17, 4
Length of a stationary object	4		35, 28, 40, 29			
Area of a mobile object	5	2, 17, 29, 4		14, 15, 18, 4		

FIGURE 2.21
First five lines on the matrix of system constraints.

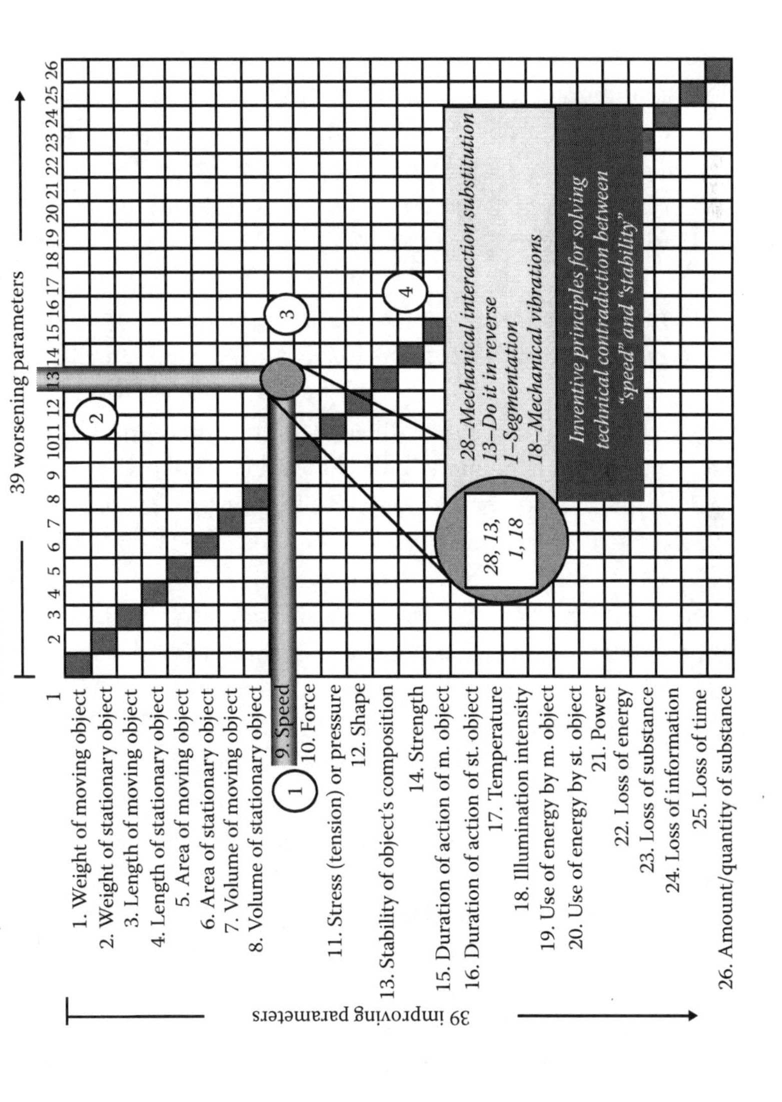

FIGURE 2.22
Small part of the TRIZ contradiction matrix. (Created by Isak Bukhman.)

points of engineering parameters for expressing technical constraints that are listed horizontally (worsening parameter) that can have a negative impact on the improvement project. At this intersection, the TRIZ principles that might offset the worsening parameter are listed. Typically, at the junction one to four, TRIZ principles are listed that may be used to offset the negative contradiction.

In the contradiction matrix, the horizontal numbers have the same description as a similar number on the left-hand (vertical) side of the matrix. For example, number 9 (SPEED) on the left-hand side of the matrix and number 9 on the horizontal top of the matrix both stand for *speed.* Number 13 on the left-hand side of the matrix and number 13 on the vertical top of the matrix are both defined as *the stability of an object's composition.*

Using the matrix to identify what should be done to offset conflict between two parameters is as simple as 1, 2, 3, and 4. (See Figure 2.23.)

- *Step 1*: Define the characteristic that you're going to change. In Figure 2.22, the characteristic that will be changed is number 9—*Speed.*
- *Step 2*: Define the worsening characteristic that will be evaluated. In this example, the worsening factor is number 11—*Stress (tension) or pressure.*
- *Step 3*: Using the matrix, determine the crossover point between number 9—*Speed* and the intersecting point with the worsening parameter number 11—*Stress (tension) or pressure.* The basic principles that are listed should be considered to offset the worsening parameter number 11—*Stress (tension) or pressure.*
- *Step 4*: At the crossover point, four of the 40 principles that most likely would offset the contradiction's impact upon the system are listed. In this case the four basic principles that are most often used to offset or minimize the worsening parameter of the matrix are:
    - Number 6—Multifunctionality
    - Number 18—Mechanical vibrations
    - Number 38—Strong oxidants
    - Number 40—Composite materials
- *Step 5*: Analyze each of the recommended 40 principles to determine how each one would impact the contradiction. Although all 40 TRIZ principles are seldom used to offset the worsening parameter, often, one or two of them lead to the right answer.

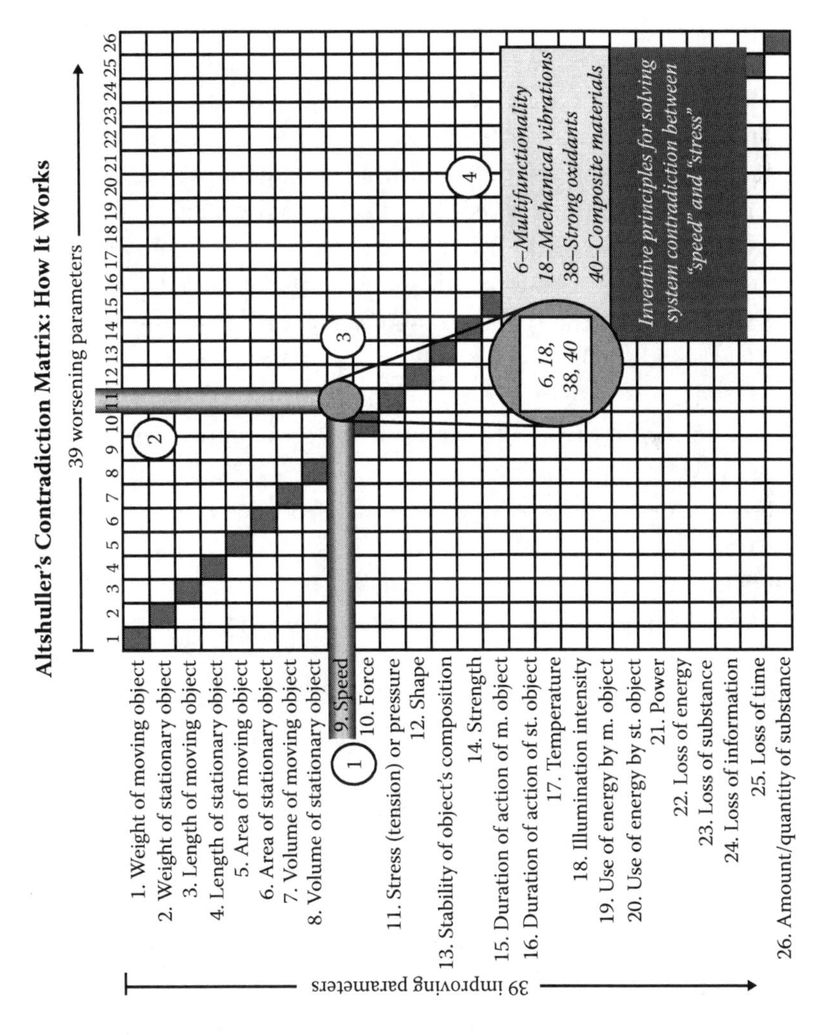

FIGURE 2.23
TRIZ Matrix. (Created by Isak Bukhman.)

Two Common Approaches to Using the TRIZ Contradiction Matrix

There are two common approaches to using the TRIZ matrix. They are as follows:

1. *Worsening parameter analysis*
   - *Step 1*: Define the improvement parameter that is considered for implementation in order to define the parameters that could result in having a negative impact upon the system.
   - *Step 2*: Review the 39 (engineering parameters for expressing technical constraints) potential worsening parameters to define the specific parameters that could relate to the individual initiative that you are working on.
   - *Step 3*: Look at the intersection between the identified worsening parameters and the proposed improvement parameter to identify TRIZ principles that could be used to offset the worsening parameter.
   - *Step 4*: Evaluate each of the recommended TRIZ principles to determine if it can be used to minimize or eliminate the impact of the worsening parameter.
   - *Step 5*: Repeat this process for each of the identified potential worsening parameters.
2. *Combined worsening parameters' analysis*
   - *Step 1*: Define the improvement parameter that you are considering to use.
   - *Step 2*: Analyze all of the recommended TRIZ principles that are in the horizontal line related to the improvement principle that is being evaluated.
   - *Step 3*: Define how many times each individual TRIZ principle is recommended for analysis across the entire horizontal line. Base this analysis on a list of how many times a specific TRIZ principle is recommended as a potential solution for the negative parameters. For example, in the horizontal line related to the improvement for *speed*, seven TRIZ principles should be evaluated to eliminate or minimize to an acceptable level the worsening parameters' impact upon the system. (See Table 2.1.)
   - *Step 4*: All other TRIZ principles are not analyzed during the initial analysis cycle. Only the ones that are recommended as a potential solution four or more times would be analyzed in this case. In other cases where there are much fewer recommended TRIZ

TABLE 2.1

Seven of the Highest-Rated Principles for Speed

Principle #	Principle	Recommended
Number 35	Transformation properties.	Recommended 9 times.
Number 38	Strong oxidants.	Recommended 10 times.
Number 13	Do it in reverse.	Recommended 6 times.
Number 19	Periodic action.	Recommended 6 times.
Number 34	Rejecting and regenerating parts.	Recommended 6 times.
Number 27	Dispose.	Recommended 5 times.
Number 28	Replacement of mechanical system.	Recommended 5 times.

principles listed, the cutoff point would be at three to four principles that are most often recommended and would be analyzed.

This could seem like a lot of work, and the calculations could be in error because the matrix is so complicated. Luckily, currently, there are computer programs that can do the same job for you in seconds, providing the user with a list of the highest potential TRIZ principles.

There are a number of ways that you can use the TRIZ contradiction matrix. Some people look at each intersection and evaluate the suggested TRIZ principles. This approach takes a great deal of time and is not in line with the time constraints we have in LTM. To make maximum use of the time, we count the number of times that an individual TRIZ principle is recommended as a potential way to offset the 16 negative contradictions that are viewed horizontally across the matrix.

- *Step 5*: Typically, the LTM team will analyze 3 to 6 of the 40 principles that are most frequently used.

Table 2.2 is a list of the number of times that the individual 40 principles as recommended as a potential principle will offset the contradiction "Energy Spent by Stationary Object." Only a few people do this by hand. There are software programs that will automatically provide you with the same relevant information, eliminating all the time required to analyze the matrix and, at the same time, are much more accurate. Using this approach as part of the LTM will result in minimizing risks and often generates new recommended improvement approaches. The end result is a much better design or process as the LTM team is not focused on eliminating waste as much as they are focused on optimizing the total system's performance.

TABLE 2.2

Analysis of Recommended TRIZ Principles for a Proposed Change of Energy Spent by Stationary Object

TRIZ Principle Recommended	Number of Times Recommended
2. Weight of a stationary object	4
10. Force	2
13. Stability of composition	4
14. Strength	1
18. Brightness	4
23. Loss of a substance	4
26. Amount of substance	3
29. Accuracy of manufacturing	3
30. Harmful factors acting on an object from outside	4
31. Harmful factors developed by an object	3
32. Manufacturability	2
37. Complexity of control	4
39. Capacity/productivity	2

Patterns of Evolution

Directed evolution is a typical advanced tool that makes use of a spider diagram to develop patterns of evolution. It seems like there are as many different approaches to using TRIZ as there are consultants available to help organizations implement TRIZ. The following is just one example of a fairly advanced concept related to the use of TRIZ theory to bring about performance improvement within an organization.

Definition: *Patterns of evolution* are a set of terms or statements that define trends that have strong, historically recurring tendencies in the development/evolution of man-made systems.

Definition: *Lines of evolution* describe in greater detail typical sequences of stages (positions on a line) that a system follows a specific pattern of evolution in the process of its natural progress. Once these positions are known, the system's current position on a line can be identified, and the possibility of transitioning to the next position can be assessed, for example, become flexible or use micro-level properties of materials utilized. Lines of evolution are grouped under one or more of the patterns of evolution that they support.

Definition: For each line of evolution, there is a group of possible corrective actions called *operators* that can be used to correct the problem or improve the product or process. An operator is a little nugget of wisdom (recommendation, suggestion) on changes to the system designed to trigger you into thinking how to solve the problem or to improve the process/product under evaluation. Operators also serve as a means to change the system position on the relevant lines of evolution (for example, using special physical effects to motivate employees).

Directed evolution has combined potential improvements into patterns of evolution. They describe in greater detail typical sequences of stages (positions on a line) that a system follows a specific pattern of evolution in the process of its natural progress. They are a set of terms or statements that define trends that have strong, historically recurring tendencies in the development/evolution of man-made systems.

There are 8 major patterns. (Some people use 12 major categories in place of these 8 patterns.) They are often presented in a spider diagram. (See Figure 2.24 for the 8 patterns of evolution and Figure 2.25 for the 12 patterns of evolution.)

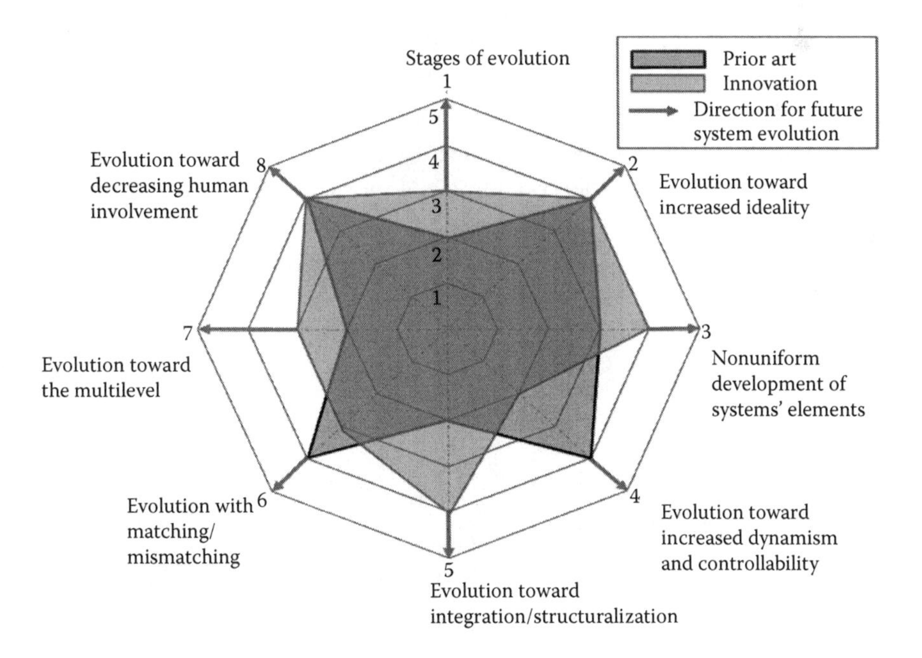

FIGURE 2.24
The 8 Patterns of Evolution.

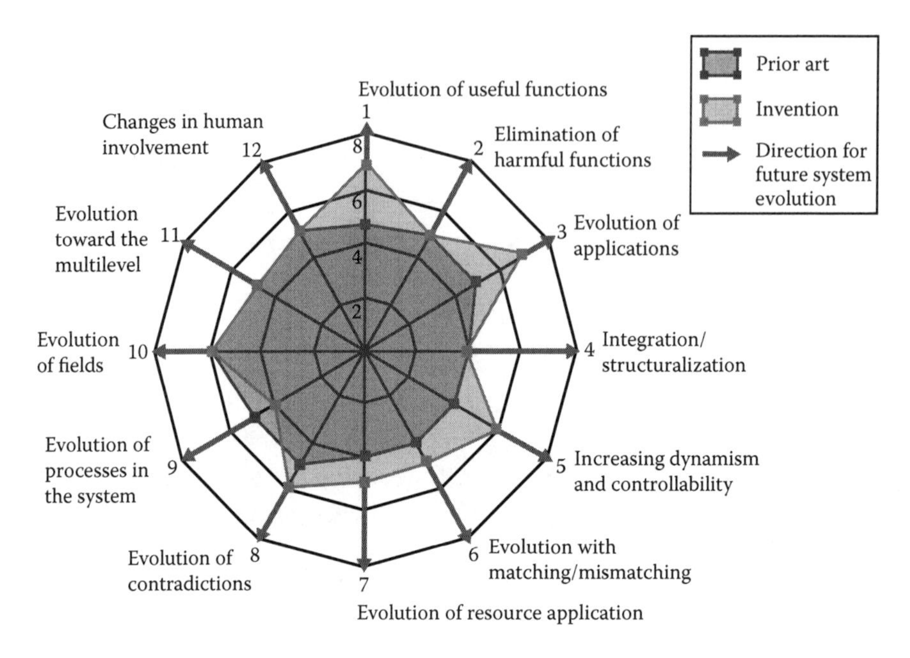

FIGURE 2.25
The 12 Patterns of Evolution.

Each of the patterns of evolution is subdivided into many lines of evolution. Lines of evolution describe in greater detail the typical sequences of stages (positions on a line) that a system follows in a specific pattern of evolution in the process of its natural progress. (See Figure 2.26.) Once the system's current position on a line can be identified, the possibility of transitioning to the next position can be assessed (for example, become flexible or use micro-level properties of materials utilized).

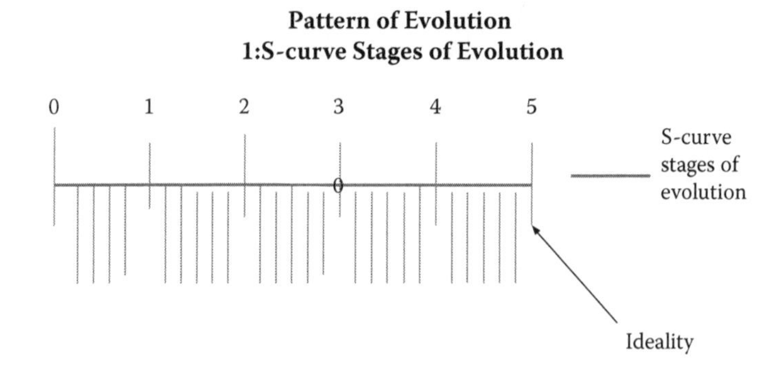

FIGURE 2.26
Lines of evolution.

Aligned with each line of evolution is a group of possible corrective actions called *operators* that can be used to correct the problem or improve the product or process. An operator is a little nugget of wisdom (recommendation, suggestion) on changes to the system designed to trigger you into thinking how to solve the problem or to improve the process/product under evaluation. They could also be used in moving the system to the next position on the line if this step requires a creative solution. Operators are drawn from the successful results of previous actions that resolved difficult technological problems and/or process problems. Operators are used to solve problems in existing systems and/or as a means to change the system position on the relevant lines of evolution, for example, by suggesting to employ the use of special physical effects. In some situations, position on the line could be described via applied operator, so a line could include a number of operators/positions that are or were applied/achieved in a certain sequence. The 40 TRIZ principles were the first operators discovered; there are over 400 of these operators used by some advanced TRIZ masters. Unfortunately, these 40 principles rest primarily in the minds of the TRIZ masters. They are not documented in a nice, convenient table for everyone to use.

The author suggests thinking of lines of evolutions as taking a long trip with many cities along the way. Each city is a destination unto itself. The operators are the Global Positioning System that tells you the best way to get to the next destination.

Note: The author doesn't want to leave you with the impression that the matrix of system constraints is the total TRIZ methodology. It is really just the starting point for the first workable tool for TRIZ that was created. There are a number of other TRIZ tools that have evolved over the last 50 years, strengthening the TRIZ methodology, computerizing it, and modifying it so that it can adequately be applied to the service environment. Some of the other tools that could be considered for use are the following:

Associate Certification TRIZ Concepts, Components, and Tools

1. **Foundational concepts**
    - Dialectics as a philosophical foundation of TRIZ
    - Directional evolution of technological systems
    - Technological system
    - Functions

- Ideal technological system
- Ideal final result
- Contradictions: administrative, engineering, and physical

Trends (laws) and subtrends (lines) of technological system evolution

- Trend of increasing degree of ideality
- Trend of nonuniform evolution of subsystems
- Trend of completeness of system parts
- Trend of *energy conductivity* of systems
- Trend of harmonization of rhythms
- Trend of transition to supersystems
- Subtrend of transition from mono- to bi- and polysystems
- Subtrend of increasing structurization of voids
- Trend of increasing dynamism
- Lines of increasing dynamism
- Trend of increasing substance–field (su-field) interactions
- Lines of evolution of su-fields
- Trend of transition from macro- to micro-levels
- Trend of matching mismatching (coordination–noncoordination)
- The general pattern of engineering systems evolution

This is just part of a methodology known as classic TRIZ. At present, TRIZ professionals around the world continue to develop and refine these concepts. For additional information related to TRIZ, we suggest that you read *TRIZ Methodology* (2012) by Isak Bukhman and/or *Innovation on Demand* (2005) by Victor Fey and Eugene Rivin.

LEAN PROCESS TRIZ

When we compare product redesign/design activities to process redesign/design activities, we immediately discover there are a lot fewer (opportunities) potential outcomes in process improvement than there are in product redesign and/or design. Product design opportunities is what classic TRIZ was created to address. This allows us to greatly simplify the classic TRIZ concepts when we apply them to process redesign/design situations. This simplified approach is called Lean Process TRIZ

(LP-TRIZ). LP-TRIZ is made up of five primary driving objectives. They are the following:

1. Cycle time–reduction improvement opportunities
2. Cost-reduction improvement opportunities
3. People-related improvement opportunities
4. Poor results or quality improvement opportunities
5. Method improvement opportunities

These five primary driving objectives can be consolidated into the following three primary improvement targets (3PIT). They are the following:

1. Cost reduction
2. Quality/reliability improvements
3. Productivity improvements

Each of these 3PIT has a number of improvement methods that, when implemented, will bring about a positive change in the related improvement targets. Figure 2.27 provides a list of the 41 process design/redesign improvement approaches.

Figure 2.27 is a list of the majority of improvement activities that are directed at improving one or more of these 3PIT. This list was created based upon the author's personal observations when he owned his own consulting firm and while he worked with Ernst & Young. While he was at Ernst & Young, Harrington had available all of Ernst & Young's knowledge management systems for the process improvement projects that Ernst & Young's consultants were involved in. Although this is not a complete list of process improvement approaches, it does include process improvement approaches that were used to improve over 95% of the projects.

Some people feel that the LP-TRIZ is even more complex than the classic TRIZ because it is made up of 41 statements, and classic TRIZ's principles are made up of 40 statements, and the engineering parameters for expressing technical contradictions are made up of only 39 statements. Just bear with us; you will soon see how simple and straightforward LP-TRIZ is.

Definitions:

- *LP-TRIZ's 41 Process Improvement Approaches (41A)*—These are 41 approaches (activities) that are most often taken to bring about improvement in process design and redesign. Each of the 41 activities

1	Eliminate operations
2	Combine operations
3	Use smaller lot sizes
4	Do away with stock
5	Eliminate all bureaucracy
6	Redo the time standards
7	Change the requirements
8	Use first in, first out
9	Perform a better risk analysis
10	Start using a sampling plan
11	Track surpluses
12	Relay out the work area
13	*Fast* dept. reorganization
14	Pick up
15	Improve sales force
16	Train people
17	Relocate people
18	Recognition
19	Prepare job descriptions
20	Put in an employee certification system
21	Get rid of layers of management
22	Work with suppliers
23	Get new suppliers
24	Supply chain management
25	Customer understanding
26	ISO 9000
27	Empowerment
28	Provide better parking for customers
29	Independent evaluation
30	Rearrange tools
31	Get rid of things not used
32	Add more equipment
33	Computerize the process
34	Better maintenance
35	Belts to help move more parts
36	Make new tools and dyes
37	Visual tracking
38	Simple language
39	Computerization
40	Online stock levels
41	Innovation

FIGURE 2.27
Major improvement approaches used to bring about a positive change in current processes.

has a positive impact upon one or more of the 3PIT. They are sometimes called The 41 Process Design/Redesign Improvement Opportunities. (See Figure 2.27.)

- *3PIT*—These are the three primary reasons that organizations redesign current processes. The 3PIT are the following:
  1. Cost reductions
  2. Productivity improvements
  3. Quality/reliability improvements
- *Five major driving objectives*—There are five major headings that each of the 41 process improvement opportunities relate to. They are the following:
  1. Cycle time–reduction improvement opportunities
  2. Cost-reduction improvement opportunities
  3. People-related improvement opportunities
  4. Poor results/reliability or quality improvement opportunities
  5. Method improvement opportunities
- *LP-TRIZ 41 × 3 improvement matrix (LP-TRIZ matrix)*—It is a column-by-column analysis of each of the 41 process improvement opportunities evaluating their performance to each of the 3PIT (cost reductions, productivity improvements, and quality/reliability improvements). All improvement activities are directed at improving one or more of these 3PIT. Figure 2.27 is a list of the 41 major ways you can improve a process or things that must be considered when designing or redesigning a process.

The LP-TRIZ 41 process design/redesign improvement opportunities may not be complete for your organization, and the author strongly recommends that you start with this basic list and add to it as your organization defines other process improvement opportunities.

After taking a closer look at the potential improvement opportunities, we soon realized that they all fit into three basic categories like a three-legged stool. (See Figure 2.28.) They are the following:

1. Productivity improvements
2. Cost reduction
3. Quality/reliability improvements

Improving processes is very similar to a three-legged stool with each leg depending on the other two for stability. These three improvement

FIGURE 2.28
Three-legged stool.

objectives are known as the 3PIT. The team should now evaluate its mission statement to determine which of the 3PIT has top priority when it comes to the process. They should also determine which of the primary improvement targets is the second most important. Of course, the one that is left is the least important. (See the following example.)

- Cost reduction = First priority
- Productivity improvements = Second priority
- Quality/reliability improvements = Last priority

All improvement activities are directed at improving one or more of these 3PIT. Keeping this in mind allows us to replace the 39 × 39 TRIZ contradiction matrix (over 1,500 potential combinations) with a 41 × 3 matrix (a total of 123 potential combinations). (See Table 2.3.)

Once we got to this point, we went out to a number of process improvement consultants and had them classify each of the 123 combinations as having a positive, a negative, or it depends on how it is used if the results are positive or negative. (See Table 2.3.)

The legend related to Table 2.3 is as follows:

- ++ (double +) = very high probability that the specific 41 process improvement approaches that are being analyzed will have a positive impact on the primary improvement target that it intersects with.
- + (single +) = good probability that the specific 41 process improvement approaches that are being analyzed will have a positive impact on the primary improvement target that it intersects with.

TABLE 2.3

LP-TRIZ 41 × 3 Improvement Matrix

	Productivity	Cost	Quality
1. Cycle-time reduction			
1.1. Eliminate operations	++	+	O
1.2. Combine operations	++	+	O
1.3. Use smaller lot sizes	N	N	+
1.4. Do away with stock	N	N	O
1.5. Eliminate all bureaucracy	++	+	O
2. Cost reduction			
2.1. Redo the time standards	+	+	NN
2.2. Change the requirements	O	+	O
2.3. Use first in, first out	N	O	O
2.4. Perform a better risk analysis	O	NN	++
2.5. Start using a sampling plan	O	+	+
2.6. Track surpluses	O	O	+
2.7. Relay out the work area	++	O	O
2.8. *Fast* department reorganization	O	+	O
2.9. Pick up	N	N	O
2.10. Improve sales force	O	O	O
3. People problems			
3.1. Train people	+	N	++
3.2. Relocate people	N	N	O
3.3. Recognition	++	O	++
3.4. Prepare job descriptions	O	O	O
3.5. Put in an employee certification system	O	N	++
3.6. Get rid of layers of management	+	O	O
4. Poor results			
4.1. Work with suppliers	O	+	++
4.2. Get new suppliers	O	+	O
4.3. Supply chain management	O	O	O
4.4. Customer understanding	O	O	+
4.5. ISO 9000	N	O	++
4.6. Empowerment	O	O	+
4.7. Provide better parking for customers	O	O	O
4.8. Independent evaluation	O	NN	+
5. Methods			
5.1. Rearrange tools	++	O	O
5.2. Get rid of things not used	+	N	O

(*Continued*)

TABLE 2.3 (CONTINUED)

LP-TRIZ 41 × 3 Improvement Matrix

	Productivity	Cost	Quality
5.3. Add more equipment	N	+	++
5.4. Computerize the process	++	+	+
5.5. Better maintenance	++	+	++
5.6. Belts to help move more parts	+	+	O
5.7. Make new tools and dyes	+	+	O
5.8. Visual tracking	++	+	O
5.9. Simple language	+	O	++
5.10. Computerization	O	N	+
5.11. Online stock levels	+	O	+
5.12. Innovation	O	+	++

- 0 (single zero) can have either a positive or negative impact based upon how it is utilized.
- N (single N) = good probability that the specific 41 process improvement approaches that are being analyzed will have a negative impact on the primary improvement target it intersects with.
- NN (double N) = very high probability that the specific 41 process improvement approaches that are being analyzed will have a negative impact on the primary improvement target it intersects with.

Each of the 41 process improvement approaches has some effect on one or more of the three primary drivers (productivity, costs, and quality). In every case, each of the 41 process improvement approaches will have a positive impact upon one or more of the three primary targets. (See +'s in Table 2.3.) In other cases, an individual process improvement approach will have a negative impact upon one or two of the other primary drivers. (In the TRIZ methodology, these results are called *contradiction*.) In Table 2.3, the negative contradictions are indicated by an *N* or *NN*. In some cases, the individual improvement approach can have a positive, a negative, or no impact depending upon how it is designed into the process. In these cases, in Table 2.3, we use a zero to indicate that the approach can have a negative, a positive, or no impact depending upon how it is applied to the process and that there is no impact on the specific combination. For example, in Table 2.3, number 2.7—*relay out the work area*—results in a very positive impact upon productivity. But both costs and quality impact are dependent upon how this is applied to the actual product, the department, and the size and weight of the equipment being relocated. For example, costs could be either positive or negative based upon how long it takes to relay out the work area and how frequently it needs to be rearranged. Quality could be either positive or negative depending on how the area is organized and what checks and balances have been added or subtracted. Another example is number 1.3—*use smaller lot sizes*. In this case, productivity is negative as time is required to reset the equipment that could have been used to produce product. Cost is negative because of this scrap that is generated during the setup of the equipment and the time it takes to set up the equipment. Quality is improved because, with the smaller lot size, quality problems related to the output are identified sooner, reducing scrap and rework. It also allows the customers' rush jobs to get through the process faster, increasing customer satisfaction.

Now, the LTM team should scan the column to find the highest-priority primary target to identify process improvement approaches that can have a positive impact upon the priority primary target. Each time that the LTM team identifies a potentially positive process improvement approach (++) for the team's first-priority primary target, they should analyze the process improvement approach to determine if it should be used to improve the process being worked on. They should start by focusing on the ++ rated process improvement approaches to estimate how it would impact the process being analyzed. (For example, in Table 2.3, if their primary focus is on improving quality, *perform a better risk analysis* would be the first approach evaluated because it is rated as ++.)

Some people find it easier to use the LP-TRIZ 41 × 3 Improvement Matrix with the zeros removed. (See Table 2.4.)

To help you use the LP-TRIZ 41 × 3 improvement matrix, we have provided the following three figures:

- Figure 2.29 shows the *productivity*-positive approaches.
- Figure 2.30 shows the *quality*-positive approaches.
- Figure 2.31 shows the *cost*-positive approaches.

Now, the LTM team should scan the column to find the highest-priority primary target to identify process improvement approaches that can have a positive impact upon the priority primary target. Each time that the LTM team identifies a potentially positive process improvement approach (++) for the team's first-priority primary target, they should analyze the process improvement approach to determine if it should be used to improve

TABLE 2.4

LP-TRIZ 41 × 3 Improvement Matrix with Zeros Removed

	Productivity	Cost	Quality
1. Cycle-time reduction			
1.1. Eliminate operations	++	+	
1.2. Combine operations	++	+	
1.3. Use smaller lot sizes	N	N	+
1.4. Do away with stock	N	N	
1.5. Eliminate all bureaucracy	++	+	
2. Cost reduction			
2.1. Redo the time standards	+	+	NN
2.2. Change the requirements		+	
2.3. Use first in, first out	N		
2.4. Perform a better risk analysis		NN	++
2.5. Start using a sampling plan		+	+
2.6. Track surpluses			+
2.7. Relay out the work area	++		
2.8. *Fast* department reorganization		+	
2.9. Pick up	N	N	
2.10. Improve sales force			
3. People problems			
3.1. Train people	+	N	++
3.2. Relocate people	N	N	
3.3. Recognition	++		++
3.4. Prepare job descriptions			
3.5. Put in an employee certification system		N	++
3.6. Get rid of layers of management	+		
4. Poor results			
4.1. Work with suppliers		+	++
4.2. Get new suppliers		+	
4.3. Supply chain management			
4.4. Customer understanding			+
4.5. ISO 9000	N		++
4.6. Empowerment			+
4.7. Provide better parking for customers			
4.8. Independent evaluation		NN	+
5. Methods			
5.1. Rearrange tools	++		
5.2.* Get rid of things not used	+	N	
5.3. Add more equipment	N	+	++
5.4. Computerize the process	++	+	+

(Continued)

TABLE 2.4 (CONTINUED)

LP-TRIZ 41 × 3 Improvement Matrix with Zeros Removed

	Productivity	Cost	Quality
5.5. Better maintenance	++	+	++
5.6. Belts to help move more parts	+	+	
5.7. Make new tools and dyes	+	+	
5.8. Visual tracking	++	+	
5.9. Simple language	+		++
5.10. Computerization		N	+
5.11. Online stock levels	+		+
5.12. Innovation		+	++

	Productivity
1. Cycle-time reduction	
1.1. Eliminate operations	++
1.2. Combine operations	++
1.5. Eliminate all bureaucracy	++
2. Cost reduction	
2.1. Redo the time standards	+
2.7. Relay out the work area	++
3. People problems	
3.1. Train people	+
3.3. Recognition	++
3.6. Get rid of layers of management	+
4. Poor results	
5. Methods	
5.1. Rearrange tools	++
5.2. Get rid of things not used	+
5.4. Computerize the process	++
5.5. Better maintenance	++
5.6. Belts to help move more parts	+
5.7. Make new tools and dyes	+
5.8. Visual tracking	++
5.9. Simple language	+
5.11. Online stock levels	+

FIGURE 2.29

Productivity-positive approaches in LP-TRIZ 41 × 3 improvement.

	Quality
1. Cycle-time reduction	
1.3. Use smaller lot sizes	+
2. Cost reduction	
2.4. Perform a better risk analysis	++
2.5. Start using a sampling plan	+
2.6. Track surpluses	+
3. People problems	
3.1. Train people	++
3.3. Recognition	++
3.5. Put in an employee certification system	++
4. Poor results	
4.1. Work with suppliers	++
4.4. Customer understanding	+
4.5. ISO 9000	++
4.6. Empowerment	+
4.8. Independent evaluation	+
5. Methods	
5.3. Add more equipment	++
5.4. Computerize the process	+
5.5. Better maintenance	++
5.9. Simple language	++
5.10. Computerization	+
5.11. Online stock levels	+
5.12. Innovation	++

FIGURE 2.30
Quality-positive approaches in LP-TRIZ 41 × 3 improvement matrix.

the process being worked on. They should start by focusing on the ++ rated process improvement approaches to estimate how it would impact the process being analyzed. (For example, in Table 2.3, if their primary focus is on improving quality, perform a better risk analysis would be the first approach evaluated because it is rated as ++.)

Figure 2.30 is a list of the process improvement approaches that have the greatest potential of improving quality within a process.

If the ++ process improvement approaches do not provide the LTM team with sufficient improvement approaches to meet their improvement

	Cost
1. Cycle-time reduction	
1.1. Eliminate operations	+
1.2. Combine operations	+
1.5. Eliminate all bureaucracy	+
2. Cost reduction	
2.1. Redo the time standards	+
2.2. Change the requirements	+
2.5. Start using a sampling plan	+
2.8. *Fast* dept. reorganization	+
3. People problems	
4. Poor results	
4.1. Work with suppliers	+
4.2. Get new suppliers	+
5. Methods	
5.3. Add more equipment	+
5.4. Computerize the process	+
5.5. Better maintenance	+
5.6. Belts to help move more parts	+
5.7. Make new tools and dyes	+
5.8. Visual tracking	+
5.12. Innovation	+

FIGURE 2.31
Cost-positive approaches in LP-TRIZ 41 × 3 improvement matrix.

goals, then they should analyze the process improvement approaches that are rated as +. The analysis of the + and ++ rated process improvement approaches usually allows the LTM team to identify the process improvement approaches that are effective in improving the primary target. In performing the analysis, the LTM team also needs to evaluate what effect the individual process improvement approach would have on the other primary targets. For example, process improvement approach number 5.1—*rearranging tools*—has a very positive effect on productivity but can have either a positive or negative effect on cost and quality depending on how it is implemented. Ideally, the process improvement approaches that are selected would have a positive impact or no impact upon the other two primary targets. This condition rarely occurs. When the selected primary target has a negative impact upon one or more of the other two primary targets (in TRIZ terminology, this is called contradictions), the LP-TRIZ team needs to evaluate how significant the negative impact would be in reducing the value-added content of the improved process.

If the negative impact is significant, the LP-TRIZ team needs to select another improvement approach that will offset that negative trend. For example, if you're trying to improve productivity and the proposed process improvement approach has a negative impact upon quality, you should refer to the LP-TRIZ matrix quality process improvement approaches and review the list to select one that will have a positive impact on quality without a negative impact on productivity. For example, the LTM team can select process improvement approach number 3.3—*recognition*—because there is a double plus (++) for productivity and a double plus (++) for quality.

The other way the LP-TRIZ matrix is used is to evaluate the results from others utilized to define improvement approaches like brainstorming and streamlining activities. At this point in the process, the LTM team has accumulated a list of potential performance improvement activities as a result of their brainstorming, streamlining, and LP-TRIZ analysis. The LTM team will analyze these potential improvements and determine which of the 41 process improvement approaches each of them is related to. Once this is accomplished, the LTM team can evaluate the proposed improvement activity using the same approaches they used in the previous paragraphs.

One word of caution—the LTM team needs to be careful so that they do not recommend improvement activities that will not add value to the end results. Frequently, one improvement approach will correct the problem or minimize it significantly. Frequently, another process improvement approach is directed at accomplishing the same results. In these cases, one of the processes adds real value to the end results, and the other approach has minimum impact upon the results after the first process improvement approach has been implemented. (For example, the team could decide to do a value analysis of the process designed to eliminate the NVA, minimize BVA, and maximize the true value. They could also decide to eliminate bureaucracy in the process.) In both cases, the LTM team is focused upon eliminating unnecessary operations. In this case, if they decide to do the value analysis of the process design, there is no need to focus on eliminating bureaucracy because the value analysis approach is also concentrated on eliminating bureaucracy.

The 41 process improvement approaches may seem a little complicated at first. We agree that it is a little difficult to point out all the nuances related to the LP-TRIZ tool, but, when it comes to implementing the process, it is very straightforward and does not contain a great deal of complexity.

We refer to this approach as LP-TRIZ because it reduces the matrix down from 39 × 39 to 41 × 3. It is also specially designed for process problems where classic TRIZ is primarily designed for new product development and/or product improvement.

SUMMARY

The five methodologies presented in this chapter are as follows:

1. VE
2. SPI
3. Lean
4. TRIZ
5. LP-TRIZ

Each of them is very important as there are many tools in each of these methodologies.

This is a chapter that you will refer back to often as you use the LTM approach.

> With the right tools, you can be an expert. Without them, you are an amateur.
>
> **H. James Harrington**

3

Starting an LTM Program

INTRODUCTION

The Lean TRIZ methodology (LTM) consists of five phases. (See Figure 3.1.) They are the following:

- Phase I—Identifying improvement opportunities
- Phase II—Preparing for the workshop
- Phase III—Conducting the workshop
- Phase IV—Implementing the change (recommendations)
- Phase V—Measuring results and rewards/recognition

Process design or redesign LTM is a methodology that is designed to bring about rapid change to processes and subprocesses, as a result of applying the following:

- Brainstorming
- Streamlining modification
- Lean approaches
- Lean Process TRIZ

Product design or redesign LTM is a methodology that is designed to help the product engineer design new products that are less expensive to build, easier to repair, and improve reliability. This improved, more robust design is the result of applying the following:

- Brainstorming
- Value engineering and analysis
- TRIZ methodology

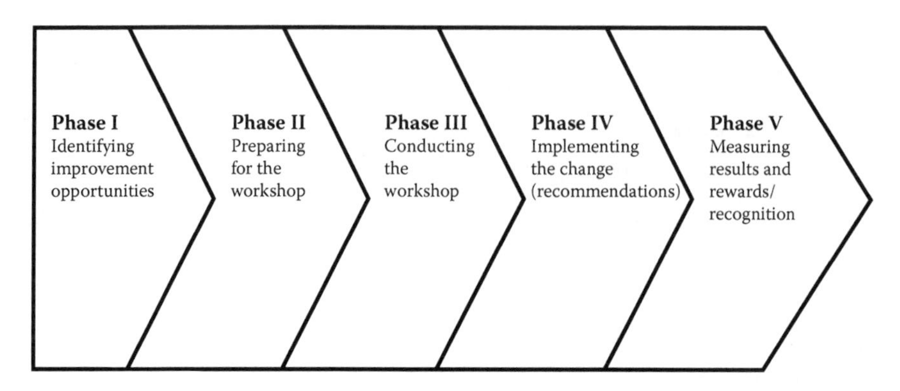

FIGURE 3.1
LTM.

In some cases, LTM will add steps to the process to facilitate its execution. Sometimes, a process contains gaps where the process breaks down. In other cases, activities will be removed as they are no-value-added activities. By simply adding a step to bridge the gap, the entire process will run more effectively. In other cases, adding an attractive feature to a present product or reducing the cost through slight changes to the design and/or the production process will open a whole new market segment to the organization.

The risks associated with short-term focus are well known. We recognize, however, that most organizations need to achieve immediate improvement, reduce costs, resolve customer problems, and improve performance on current products, which, in turn, funds long-term improvement initiatives. It is frequently necessary to sustain market share on the current product until a radical new design has completed the research and development phase and the production phase and reestablished itself as a viable alternative to the present product. This usually occurs when the new product becomes the preferred product by delivering more value to its customers. As a result, an approach like LTM is ideally suited to be part of a portfolio of improvement initiatives designed to address those areas that will provide your organization with the greatest immediate value.

LTM is a simple technique as opposed to the more complex, time-consuming techniques used in traditional TRIZ, process redesign, process reengineering, process benchmarking, and Six Sigma. LTM is used to identify and develop short-term improvement initiatives.

SELLING LTM TO YOUR ORGANIZATION

The first hurdle that must be faced when implementing LTM in your organization is selling the concept to management. Like any new project, this first requires a champion. A champion is someone who believes in the LTM process to the point that he or she wants, at a minimum, to see an LTM program implemented. We hope that, by the time you finish reading this book, you will become that champion, if someone else has not already stepped up to fulfill that role.

The champion needs to build up a level of interest and commitment so that the pilot program can be approved. Often, the person who discovers a new, excellent approach is not a manager or at a management level that is high enough to approve a major project. In this case, he or she will need to build interest in the appropriate managers who can lead to the approval of the project. LTM is easy to sell as it just requires a small investment for a big, fast payback.

Frequently, the champion will have a consulting firm come in and make an LTM presentation to the senior management to get their buy-in and support.

We suggest that the organization starts the LTM by doing a pilot of two or three two-day projects. The LTM pilot program should relate to the organization's strategic plan. We find that it is best to start the LTM project using an already trained and experienced LTM facilitator. The concept is simple, but, during the start-up, this facilitator needs to have credibility and needs to build awareness, sponsorship, and an LTM value proposition. Having an experienced facilitator who has been involved in LTM projects before greatly reduces the risk of failure while maximizing the use of the other resources. Consulting firms like Ideation International, Technical Innovation Center, and Harrington Management Systems all have experienced TRIZ facilitators who can fulfill this need. The initial presentation to the management team should identify current products that need to evolve to the next level, processes that they would like to improve, or improvement opportunities that they would like to take advantage of by applying LTM. In addition, the management team should be invited to define processes/subprocesses and/or products that they believe would be good candidates and that they would volunteer to play the sponsor's role for.

LTM KEY ROLES AND RESPONSIBILITIES

It is important that the management team be made aware of the roles and responsibilities that various individuals within the organization will need to assume if and when the LTM is widely accepted. The LTM program will involve many different people, each performing different roles. Throughout this book, we will be using the following titles and corresponding responsibilities for each of these roles. A detailed description of these roles will be presented in the appropriate parts in the book.

- *LTM champion*—This individual is the person who is in charge of the LTM program for the organization. He or she should have a good understanding of classical TRIZ, streamlined process improvement, and value engineering. We recommend that he or she be a certified TRIZ practitioner, a business process improvement practitioner, and a Lean Six-Sigma Black Belt. Individuals who are certified at the Associate or Green Belt levels usually do not have the experience required to handle an LTM program. The LTM champion is the individual who is assigned the responsibilities to maintain an LTM program throughout the organization.
- *LTM change champion*—This is the individual who is responsible for coordinating the day-to-day implementation of the recommendation(s). The individual who is assigned this responsibility should be a member of the workshop team. The change champion is the one who makes it happen.
- *LTM change sponsor*—This is the individual who is responsible for providing resources and handling roadblocks that the LTM team leader can't handle. He or she is responsible for approving the suggestions that will be implemented. He or she sponsors individual changes by providing resources and measures performance throughout the implementation. The change sponsor is accountable to the LTM sponsor for implementation. During the implementation process, the change champion may face challenges such as conflicting priorities that he or she is unable to resolve. The change sponsor is then responsible for resolving these issues. The change sponsor may need to be assigned during the presentations to the LTM sponsor or immediately following the meeting.

- *LTM sponsor(s)*—This is the individual who approves an individual LTM project and decides which changes are implemented. This is a manager who is at a high-enough level that he or she can authorize changes to the process or product without higher-level management approval. The sponsor will attend the last part of the workshop where the LTM team members present their recommended changes to him or her. The sponsor is committed to make a decision at this time on each suggestion. The sponsor must either accept or reject the proposal; he or she does not have the option to put it off to a later date. This firm, on-the-spot decision-making process is a critical part of the LTM process.
- *LTM facilitator*—This is the individual who has a detailed understanding of the LTM, how to use its many tools, and how to manage a team meeting. He or she helps the project LTM team leader keep the workshop on schedule so that it meets the desired goals. It is helpful if this individual is a certified classical TRIZ practitioner and has a good understanding of classical TRIZ, organizational change management, and streamlined process improvement. Usually, he or she is assigned to two to four LTM projects. He or she will advise the LTM leader related to the proper tools they should use under the present situation to get maximum results.
- *LTM team leader*—This is an individual who is assigned to lead an individual LTM Team. He or she schedules meetings, manages the meeting, and follows up on specific assignments.
- *LTM team members*—These are the individuals who attend the LTM workshop and create the future-state solutions.
- *LTM coordinator*—In big organizations, when they get many LTM projects going on at the same time, the LTM champion may need someone to help manage the portfolio of projects. This individual is referred to as the LTM coordinator and should have an understanding of the LTM.

LTM Facilitator/Leader and Team Members' Responsibilities

The LTM facilitator plays a key role in the LTM process. He or she serves as the LTM process expert (consultant). Facilitation is defined as the complex skill that motivates, enables, and empowers a group of people to complete a task. To be a facilitator is to act as a human catalyst who can turn a collection of separate individuals into a working team.

The facilitator plays a very specific role in the workshop. The facilitator's job is to

- Focus on how well people work together to ensure that the members of the group can accomplish their goals
- Equalize the responsibility for the success or failure of the group according to its defined goals and functions
- Allow more people to have control in determining what happens in the group and what decisions are made
- Build team effort on solid foundation of data
- Help the leader do the job
- Measure and monitor team goals and objectives
- Provide feedback

The facilitator's role can be divided into four quadrants. They are the following:

- *First quadrant—Prevention (25%)*
  - Build agendas.
  - Build detailed TRIZ process road maps.
  - Identify issues in the large system that affect the meeting.
  - Identify strategies to handle issues and problems.
  - Think strategically.
- *Second quadrant—Theory and techniques (25%)*
  - Understand the conceptual LTM road map of problem-solving and LTM.
  - Has a repository of LTM process suggestions and tools available.
  - Know how to introduce the LTM process suggestions and tools.
  - Understand the roles, attitudes, and behaviors of the team and the theory that underlines them.
  - Understand organizational change management.
- *Third quadrant—Process awareness (25%)*
  - Know the difference between process and content.
  - Recognize strategic movement.
  - Know what phase of LTM activities are in.
  - Able to see and describe the design/process that the team is assigned to improve or redesign.
  - Understands which technique is best suited to solve specific design/process.

    - He or she will need to be able to describe the different methodologies to the LTM team so that they will be able to use them under his or her guidance.
- *Fourth quadrant—Facilitation/actually doing it (25%)*
    - Make LTM process observations.
    - Make LTM process suggestions.
    - Introduce and provide training on the LTM tools.
    - Implement interventions.
    - Get agreement.
    - Facilitate nonverbally.
    - Remain neutral.

Figure 3.2 compares the team roles and responsibilities for the team TRIZ facilitator, the team leader, and the team member.

Figure 3.3 lists the detailed roles and responsibilities for the team facilitator.

Figure 3.4 lists the detailed roles and responsibilities for the team leader.

Figure 3.5 lists the detailed roles and responsibilities for the team member.

	Team facilitator	**Team leader**	**Team member**
Purpose	To promote effective group dynamics	To guide teams to achieve successful outcomes	To share knowledge and expertise
Major concern	*How* decisions are made	*What* decisions are made	*What* decisions are made
Principal responsibilities	• Ensure equal participation by team members • Mediate and resolve conflicts • Provide feedback and support team leaders • Suggest problem-solving tools and techniques • Provide training	• Conduct team meetings • Provide direction and focus to team activities • Ensure productive use of team members' time • Represent team to management • Document team activities and outcomes	• Offer perspective and ideas • Participate actively in team meetings • Adhere to meeting ground rules • Perform assignments on time • Support implementation of recommendations • Collect data • Develop corrective action plan • Estimate savings
Position type	Organization-wide	Team specific	Team specific
Selection criteria	Personal characteristics/process skills/TRIZ skills	Leadership skills/TRIZ skills	Process understanding as critical skills

FIGURE 3.2
Team roles and responsibilities.

	Facilitator roles and responsibilities
Purpose	To promote effective group dynamics and facilitate the process by which teams engage in problem-solving
	Concerned with *how* decisions are made
Principal responsibilities	Seek opinions of team members, synthesize different ideas, test consensus, and summarize key points during team meetings
	Mediate and resolve conflicts among team members
	Elicit opinions of less vocal members and make certain that more vocal members do not dominate team meetings
	Assist the team to achieve its goal by creating balance between individual and collective capabilities
	Ensure that appropriate documentation of team meetings is maintained
	Provide feedback and direction to team leaders on their ability to manage both team process and outcomes
	Suggest LTM tools and techniques to assist team members in problem-solving
	Assist team leader in training team members on use of problem-solving tools and techniques
	Conduct training and guide the development of team leaders
	Arrange for internal consulting assistance as necessary
	Periodically update LTM champion on team functioning and progress
	Time boxes in the items of the agenda
Required knowledge and skills	Knowledge of LTM improvement concepts, tools, and techniques
	Strong group dynamics and intervention skills
	Understands and applies analytical thinking
	Knowledge of basic behavioral science theory
	Experienced in instructing adult learners
	Effective at group presentations
	Knowledge of organizational change methodology
Personal characteristics	People oriented
	Good listener and communicator
	Positive, confident attitude
	Strong personal commitment to performance improvement

FIGURE 3.3
Facilitator roles and responsibilities.

The LTM facilitator's activities differ from the way a standard facilitator handles other meetings for a number of reasons. For example,

- Tight time frames need to be observed. The LTM process moves very quickly. As a result, directive facilitation techniques may be necessary.
  - Requires greater flexibility in approaches in driving toward solutions.
- Business content is value added to teams.
  - Share best practices and perceptions.

	Leader roles and responsibilities
Purpose	To guide teams in achieving successful outcomes through LTM-structured problem-solving and implementation of recommendations; concerned with what decisions are made
Principal responsibilities	Conduct team meetings, provide direction and focus to team activities, and assess team progress
	Guide the team without dominating it, promote free exchange of ideas, and involve all members in the problem-solving process
	Use effective meeting skills and focus on tasks, time allocation, and work methods to ensure time is used productively
	Encourage creative problem-solving and guard against innovation killers
	Instruct team members in use of LTM tools and techniques
	Represent the team to management
	Interface with other teams and support resources
	Coordinate arrangements and administrative details for team meetings
	Ensure that proper documentation of process and outcomes is maintained for all team meetings and activities
	Assist in tracking and measuring progress on team activities
Required knowledge and skills	Familiar with improvement LTM and other problem-solving concepts, tools, and techniques
	Knowledge of the products/process being studied
	Use of effective meeting skills
Personal characteristics	Positive attitude
	Able to work with individuals and teams
	Not domineering, critical, and bossy
	Good listener
	Trust and support others
	Share rewards and recognition

FIGURE 3.4
Team leader roles and responsibilities.

- Coaching skills are critical.
  - Range from sponsor relationships to team members.
- Must be skilled in driving teams to consensus very quickly so that there are no surprises during the presentation to the sponsor.

Internal LTM Facilitators

Many organizations want to manage their own LTM workshops. If this is the case, the following is a typical LTM facilitator training cycle:

- *Basic facilitator training (three to five days).* This is basic training related to how to facilitate a team to get maximum use of the creativity and time allotted for the team's meeting.

	Member roles and responsibilities
Purpose	To share knowledge and expertise of the process being studied and to participate fully in the problem-solving process
Principal responsibilities	Offer perspective and ideas on issues already addressed by the team
	Participate in all team meetings, discussions, decisions, and activities
	Recognize that team participation is part of one's *real job*
	Adhere to meeting ground rules established by the team
	Perform assignments between meetings
	Assist in preparing and presenting documentation and reports on team activities and outcomes
	Serve as timekeeper or recorder of minutes as requested by team
	Take part in setting goals and developing action plan for team
	Recommend agenda items for future meetings
	Critique and offer suggestions for improving the meeting process
	Implement recommendations of team and monitor results
	Evaluate/estimate the impact of suggested improvements
Required knowledge and skills	Work with some aspect of process being studied
	Familiar with problem-solving tools
Personal characteristics	Positive attitude
	Able to work with individuals and teams
	Respect and trust others
	Share rewards and recognition

FIGURE 3.5
Team member roles and responsibilities.

- *Facilitator LTM training (two days).* The focus of this training is on the LTM process and how to implement the process within an organization.
- *Attend an LTM workshop.*
- *Observe an entire LTM process.*
- *Cofacilitate an entire LTM process (one to three times).*
- *Take the lead role on an entire LTM process.*

A facilitator would typically do three to four workshops per month if they are working full-time on the LTM program. All workshops should be team facilitated with one or two facilitators for each workshop.

Preparation and Follow-Up

Each LTM project requires effort to prepare for the workshop. This effort includes going on and collecting functional information that will allow the team to do effective root cause analysis. Following the workshop, effort is required to follow up to ensure that the approved suggestions are implemented and that value improvement objectives are met. This is usually a two-man job that includes the facilitator and the team leader.

SUMMARY

We don't see LTM as a stand-alone improvement process. It works best as a part of an overall focus on organizational improvement using methodologies like TRIZ, business process improvement, Lean, total quality management, or total improvement management. It effectively supports methodologies like process redesign or reengineering. LTM is often used to start major improvement methodologies as it provides fast, positive feedback to management and the employees. It also helps to free up resources to man the other improvement methodologies.

Most LTM projects are measured by the dollars they save the organization. An average LTM project in a large organization saves approximately $250,000 per project (not bad for a two-day investment). But we prefer to look at measuring the LTM program's impact upon quality, customer satisfaction, increased market share, the extended life of a basic product, reduced product cost, and reduced cycle time. Most of these measurements result in making or saving money, but, unfortunately, it is hard to convert these soft measurements into the hard reality of dollars saved or generated. It is difficult to convert increased customer satisfaction or increased customer loyalty into dollars generated unless your financial organization is advanced in its thinking and operations. Saving money is important, but it is not everything.

As good as LTM is and as easy as it is to implement, there are usually some risks related to applying it to an organization. Typical risks are the following:

- How will the organization manage the discomfort with imprecise matrix?
- How will the organization manage the discomfort with immediate decision making?
- How will the employees' resistance to LTM be managed? Will it be looked at as a *flavor-of-the-month* improvement package?

> The only people who won't like LTM is your competition when they see results.
>
> **H. James Harrington**

4

Phase I—Identifying the Opportunities

Problems are just opportunities that have not been worked on yet.

H. James Harrington

INTRODUCTION

As you can see in Figure 4.1, Phase I—Identifying the opportunities is the starting point in the Lean TRIZ methodology (LTM). It leads directly into Phase II—Preparing for the workshop.

Phase I of the LTM is *identifying the opportunities*. Five activities take place during this phase. They are the following:

- Activity 1: Identifying the need
- Activity 2: Using the S-curve
- Activity 3: Other ways to identify LTM projects
- Activity 4: Screening and prioritizing potential LTM projects

Identifying an LTM opportunity can be done by anyone in the organization. These opportunities are typically processes or subprocesses that have too many delays, cost too much, have too much bureaucracy built into them, and/or don't meet customer requirements. The LTM is most frequently used to make changes to an already-established product that has saturated the market and/or whose competition is eroding the organization's share of the market.

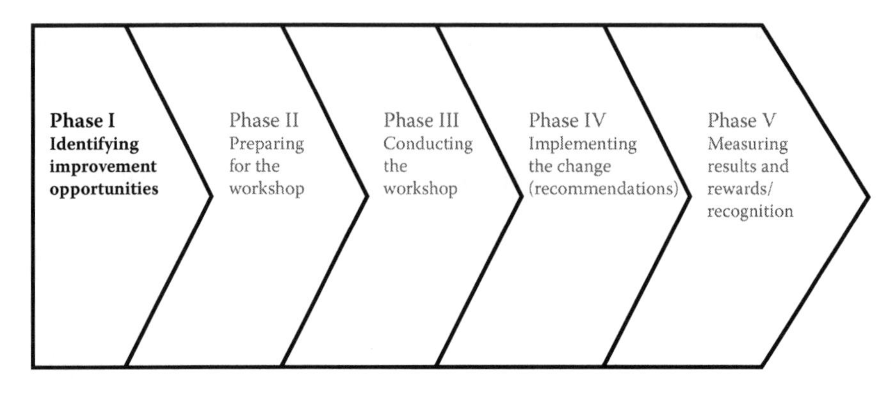

FIGURE 4.1
Phase I—Identifying the opportunities.

As we pointed out earlier, LTM is not designed for that 5% of the very complex problems or design. Misapplying it can result in loss in confidence in the LTM. The identification and timing of the LTM opportunities are a crucial step in completing the cycle successfully. The author wants to emphasize that both the identification of the correct type of improvement opportunity and the timing of when the new system can be implemented are absolutely critical to the success or failure of the project. Selecting a project that is overly complicated for the LTM will provide the organization with less than an optimum solution.

ACTIVITY 1: IDENTIFYING THE NEED

There are many ways that improvement opportunities can be identified. Some of them are the following:

- Customers' complaints or suggestions
- The result of focus groups
- Decreasing profits
- High costs to deliver service or product
- Saturation of current market
- Safety-related problems
- Employee suggestions
- Advances in technology

- Supplier suggestions
- Competitive advances
- Long cycle times

All organizations use processes to organize their activities in order to produce outputs. Unfortunately, many of these processes grew out of need, not out of design. The result is they have become overburdened with no-value-added activities that are costly. We live with too much bureaucracy in our processes. Why? It is because they are working, so why fix it if it is not broken? And, when it does break, we jump in and add a quick fix to get it up and running again. Our processes look more like patchwork quilts than design processes. The processes have become overburdened with checks, approvals, revisions, delays, rework, and needless movement.

It is obvious that there are many opportunities for improvement. Too often, we have addressed this waste by adding in computer programs that just add additional no-value-added cost to the total operation but do not streamline the process. In other cases, the new software packages are designed to give the organization a competitive advantage, but, in truth, that is not what happens. The organization spends thousands of dollars to install and operate the new software packages, but the organization's competitor is implementing the same software packages. As a result, the new software packages do not provide the organization with a competitive advantage. Installing the latest software packages usually does not provide the organization with a competitive advantage, but, if they were not installed, the organization would have a major competitive disadvantage. All that really occurs is that installing the software packages just results in increasing overhead costs. Yes, you do need the latest software packages installed in your organization, but installing them is not going to give you a competitive advantage. However, not installing them can give you a major competitive disadvantage.

The average organization has hundreds, if not thousands, of processes that have three to eight subprocesses included in each one. Most of these subprocesses have waste built into them. Many of these subprocesses require more of a mind-set and associated simple process changes rather than significant investment in new technology or information systems. In many cases, an evolutionary product design based on the already-established product is an excellent way to offset declining profits. To capitalize on these quick-win opportunities, LTM was developed to help organizations to achieve immediate impact and value. LTM is a fast-paced, high-focused process improvement approach that is designed to

- Streamline subprocesses, adding value by slight modifications to the current process
- Attack and eliminate the bureaucracy that is built into processes
- Bring about rapid change within the organization
- Quickly take advantage of opportunities for immediate improvement
- Tap the ideas and develop commitment to the improvement for the employees involved in the process
- Design evolutionary changes to present products
- Move an obvious improvement up to minor or major improvements
- Provide the organization with an approach where they can create solutions and new creative products on demand

Once you have completed the pilot, the organization should have enough positive results to justify expanding LTM to other parts of the organization. In this chapter, we will discuss some of the ways candidates for LTM surface, are screened, and then are selected.

It is extremely important that Phase I be very effective. If the wrong processes or products are selected, the team may not be able to complete the workshop in the two days allotted, or the corrections that are needed cannot be completed within 30 days after the workshop. Select design products and design problems that are evolutionary rather than revolutionary.

In summary, anyone can recognize the need for an LTM project. All they have to do is to realize that a specific product, process, or department is as follows:

- Too costly.
- Sales are dropping off.
- Competition has come up with more desirable features.
- Has too much bureaucracy.
- Products that reach the top of the S-curve.
- Takes too long.
- Requires too much movement.
- Has quality problems.
- Is hard to accomplish.
- Has too many delays.
- Is not effective.

- Is not meeting customer requirements.
- Is not meeting efficiency requirements.
- Has potential to be better than it is.

Based upon our experience, most LTM projects result from suggestions made by the management team. Once the pilot project is successful, you are often flooded with requests from all levels of management.

Communication

If the organization wants everyone to submit recommendations for LTM projects, the organization needs to get the LTM message out to the masses. The best method to do that is to post on the bulletin board and publish in the organization's newsletter the result of the pilot projects and include an invitation to submit a suggestion for an LTM project. Typically, these suggestions are submitted to the LTM champion or, if the organization has a formal suggestion program, to the suggestion department, which then forwards them to the LTM champion. Often, when the LTM champion is a manager, he or she doesn't have the time to coordinate the LTM program. As a result, an LTM coordinator is appointed to do the day-to-day management of the LTM program. We suggest that, in each newsletter, you feature at least one of the LTM projects, keeping the LTM program visible to everyone. It also provides very good recognition to the LTM team members, particularly if a picture of the team is included. We also suggest that, once a month, the results of the LTM program should be presented to the executive committee by the LTM champion. Often, at these meetings, additional LTM projects are identified.

Just because an item is suggested for the LTM program does not mean that it will be accepted. Each suggested project needs to have a present value calculation completed for the projection of potential savings and a sponsor for the suggested project.

ACTIVITY 2: USING THE S-CURVE

The best way we have found to identify potential LTM opportunities is by analyzing the individual product's S-curve. The S-curve portrays the general pattern between a new product and customer demand. Most successful organizations start out with a new product with only a small initial group of customers. This is followed by a period of rapid growth as the general public starts to seek out the new product. Eventually, the demand for the product peeks out and levels off as the market matures. As the demand for the product is fulfilled, sales drop off. To offset this downward trend, organizations frequently start offering new features or reduce prices to the customer base. When the S-curve starts to drop off, it is time for the organization to introduce a replacement product or service or to come up with evolutionary changes in the current product to stimulate current and new customers to procure the product or service. Figure 4.2 shows the evolution of a product that has no changes made to it.

During Stage 0, the product or service is still in the research and development (R&D) phase. It may or may not complete the cycle and be delivered to an external customer. This is the very costly part of the cycle as money is being invested and little money is being generated. Basically, the product does not exist, but it has the potential for becoming a legitimate project by the end of Stage 0.

During Stage 1, the product has the potential of being a deliverable product or service to the external customer. Work is going on to obtain customer feedback and set up the facilities that will be delivering products or services to an external customer. This is also the stage where the cost of

Evolutionary positioning: S-curve analysis

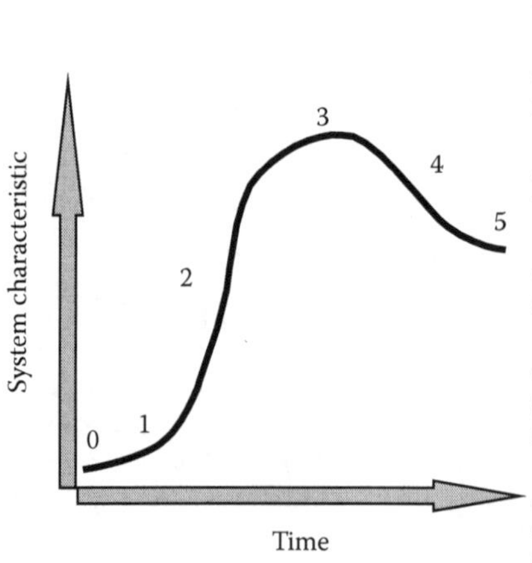

Stage 0—A system does not exist yet, but important conditions for its emergence are developing.

Stage 1—A new system appears due to a high-level invention and begins developing slowly.

Stage 2—Begins when society recognizes the value of the new system.

Stage 3—Begins when the resources on which the original system is based are mostly exhausted.

Stage 4—Begins when a new system (or the next generation of the current system) emerges to replace the existing one.

Stage 5—Begins if the new system does not completely replace the existing system, which still has limited application.

FIGURE 4.2
Typical S-curve.

activities far exceeds any revenue that is generated. By the end of Stage 1, the value proposition for the product or service has been prepared, and the manufacturing, marketing, and sales are charging ahead as customers begin to recognize the value of the new product or service. Sometimes, during Stage 2, the income generated exceeds the cost of providing the product or service, and the net losses during Stage 0 and Stage 1 are beginning to be offset.

During Stage 3, market grows as customer demand increases rapidly. Sustaining the product and/or service at the peak of Stage 3 is the object of successful sales and marketing. It is at the point in the S-curve where maximum profit from the project is realized. The objective of the organization is to stay at the Stage 3 level as long as possible or to introduce an evolutionary product that will increase the size of the total market. Normally, it is during Stage 3 that the replacement or upgraded product starts into Stage 0 in order to reach Stage 2 prior to the initial product starting its downward trend (Stage 4).

During Stage 4, the demand for the product or services is dropping very fast, and the organization needs to be able to reduce the sales price or provide a product or service that adds new and increased value to the customer in order to turn the negative trend around. This can be accomplished by producing and providing increased features that restimulate the market growth (an evolutionary product or service solution). Most of the advancements made in products and services are evolutionary solutions as they require little or no R&D activities. Often, the evolutionary design has very little impact upon how the product or service is produced. For example, the auto industry's yearly model change is primarily an evolutionary improvement where there is a slight change to the body styling in order to stimulate a demand for the current year's models. Figures 4.3 and 4.4 demonstrate the impact of evolutionary solutions for an already-established product or service.

The first curve on the left-hand side in Figure 4.3 represents the impact a radical new design has on the market. The other curve shows the impact of an evolutionary cycle. Evolutionary improvements can maintain and often grow the organization's market share. In most cases, the timing for the start of the design of an evolutionary product or service needs to start early in Stage 3 so that the evolutionary design is ready to be delivered to the customer prior to the start of market turndown that is the characteristic of Stage 4. One of the major objectives we have in LTM is to start an LTM project soon after the product/service has entered into Stage 3.

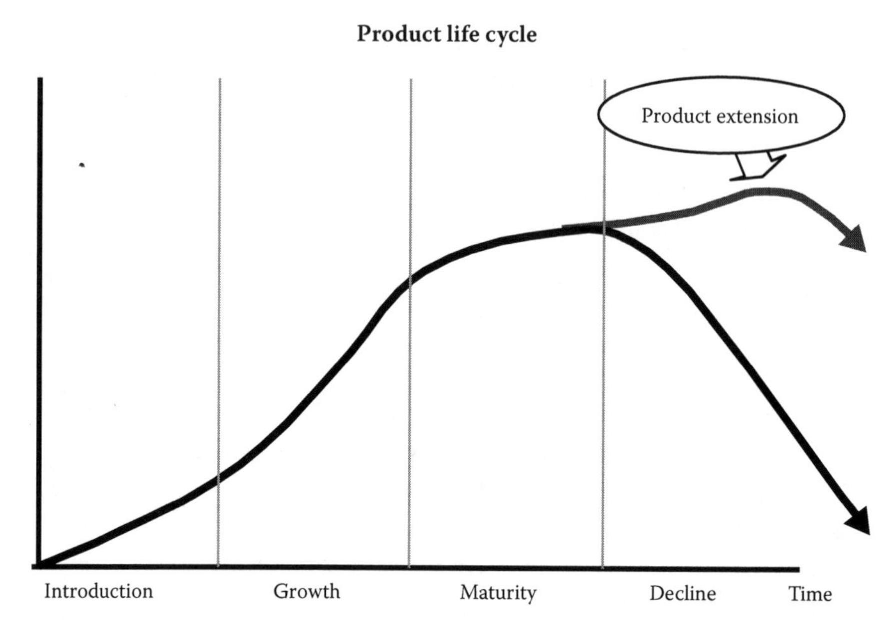

FIGURE 4.3
Diagram of the sales performance of a product or service that cycle through one evolutionary improvement of the same basic product or service design.

High performance: The climbing and jumping of S-curves

In the world of innovation, an S-curve explains the common evolution of a successful new technology or product. At first, early adopters provide the momentum behind uptake. A steep ascent follows, as the masses swiftly catch up. Finally, the curve levels off sharply, as the adoption approaches saturation. For any new and successful business, the process is much the same for the sales of its products and services. High performance is defined by companies that execute repeated climbs and jumps of the S-curve.

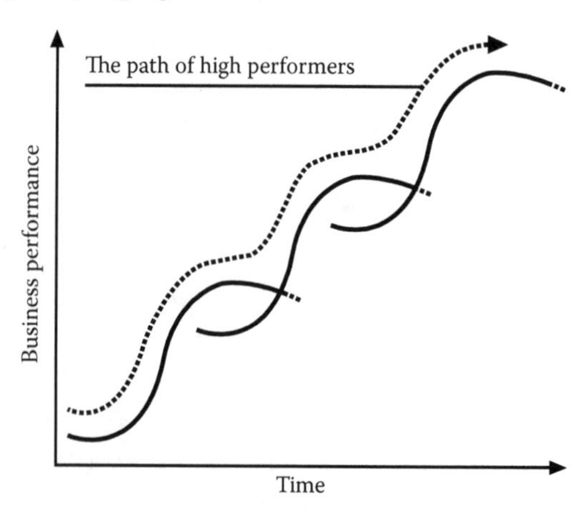

FIGURE 4.4
Diagram of the sales performance of a product or service that cycle through two evolutionary improvements for the same basic product. (From Acenture, 2010.)

Figure 4.4 illustrates two evolutionary design cycles that not only sustained the present market share but also grew the market.

At some point, in almost all product cycles, evolutionary improvements are no longer able to sustain the product's value as the present product market has been saturated or when a radically new design goes on sale and the old product becomes obsolete. We do not recommend LTM as a process to develop a revolutionary product or service. When a radically new product or service is required, we recommend the use of classical or advanced TRIZ applications.

ACTIVITY 3: OTHER WAYS TO IDENTIFY LTM PROJECTS

Customer feedback is one of the best ways to trigger an LTM project. The following are some of the most effective methods to collect customer feedback:

- *Surveys*

  Customer surveys are one of the most effective ways of collecting large amounts of information related to your product or service. Ask these questions to get the best and most effective survey data: "How can I improve that? What did I do wrong?"
- *Suggestions*

  The suggestion system in the organization is full of ideas to correct problems and improve performance. Most of them attack the problem in a superficial patchwork way. By analyzing the suggestions related to the different areas or processes, this often highlights the places where an LTM project is needed.
- *Strategic planning*

  The strategic plan is another fruitful place to look for defining LTM projects. A lot of hard work, knowledge, and experience went into preparing these very valuable plans. The LTM champion should study them to define where the priority opportunities are and direct the LTM projects to support these strategic plans.
- *Key performance indicators*

  Key performance indicators are another place to look in order to define LTM projects. They are the critical measurements for the organization. By studying them, the LTM champion can define the processes that drive these measurements. Once these major processes are defined, the LTM champion should analyze them to define products and/or subprocesses that are good candidates for LTM projects.

- *Department performance indicators*

  Department performance indicators (DPIs) are the measurements of the major processes that go on within the departments or at the natural work team levels. In this case, we define a major process as one that uses 10% or more of the department's resources. Each process usually has two or more DPIs related to it with a target for each measurement. (Typically, there are two to five major processes at the department level.) By reviewing these measurements, the LTM champion can quickly define the departments that are having process and product problems. But even this can be misleading. In addition to defining what targets have not been met, look at the long-term aspects related to the department. At least three years of performance should be analyzed to determine if there is at least 8% improvement in each measurement every year. If not, then these department processes may also be LTM project targets.

- *LTM workshops*

  Often, during an LTM workshop, the team members recognize that there are other applications to apply the LTM to. These ideas, although not part of the current LTM project, are recorded on the parking board. (A parking board is a whiteboard/flip chart that is used to record ideas that are not in direct line with the subject being discussed but will be handled at a later date.) These are usually very good suggestions because the team is very familiar with the LTM, and they can readily see how it can be applied to other situations.

- *Focus groups*

  Focus groups have been proven to be one of the most effective ways of probing into future customer needs and opportunities for improvement in current products. A focus group will typically be made up of 8–14 customers or potential customers. To make the best use of the customer's time, an experienced facilitator should probe into a predetermined list of subjects. He or she must be effective at allowing the exchange to drift off the specific list of questions.

ACTIVITY 4: SCREENING AND PRIORITIZING POTENTIAL LTM PROJECTS

It is very important that the LTM champion screens the suggested LTM projects to be sure that they fall into the workshop format. The LTM champion should consider the following criteria:

- Can the situation be addressed adequately in 8–12 hours of problem solving?
- Will the solution require more than 60 days to implement? (Many of the information technology solutions are out of scope for LTM projects.)
- Will management be able to approve the suggestion?
- Are there too many departments involved in the situation?
- Will a 10% to 20% improvement be adequate?
- Is the situation better suited for one of the other methodologies (value engineering analysis; activity-based costing; reengineering; redesign; or define, measure, analyze, improve, and control?

Large, cross-functional processes suggested for LTM and having major improvement objectives for the workshop should be examined for ways to divide the large process into smaller, isolated pieces where workshops could be scheduled for each individual item.

Prioritization

If there are a lot of suggested LTM projects and resources are limited, it will mean that the LTM projects will need to be prioritized. In prioritizing the LTM projects, the following factors should be considered:

- *Customer impact*—How much does the customer care?
- *Changeability index*—Can it be fixed?
- *Performance status*—How broke is it?
- *Business impact*—How important is it to the business?
- *Work impact*—Are the resources available?
- *Sponsor support*—Is there a sponsor available who is very supportive of the process?
- *Value analysis*—How much value does it add to the organization?

There are three common ways used to prioritize LTM projects. They are the following:

1. Management selection approach
2. Two-level analysis
3. Weighted selection approach

Management Selection Approach

In this approach, the list of potential LTM projects are prepared by the LTM champion or facilitator and discussed with a group of executives/potential sponsors. A typical list would include 20 potential LTM projects. After discussing the pros and cons of each project, the executives will write down on a card their first, second, third, and fourth choice. The LTM champion will record the results of each person's evaluation on a whiteboard or a flip chart rating each input as follows:

- Rating 1 equals four points.
- Rating 2 equals three points.
- Rating 3 equals two points.
- Rating 4 equals one point.

The LTM projects with the highest point scores are the ones that are scheduled for LTM workshops. Don't let the point scores be the only guide. Don't let it override good management judgment.

Two-Level Analysis

Another approach to prioritization is selecting two key factors and comparing them to each other. Figure 4.5 compares improvement opportunity versus impact upon the business unit's performance. In this case, each LTM opportunity is evaluated. For example, if the improvement opportunity is rated as being low and the impact on the business is rated as being low, then this opportunity would fall into the first quadrant and probably not be a good candidate for the LTM. Those options that are rated high in both factors are the top priorities. This is not as self-serving as it might seem. As you improve the internal workings of an organization, you reduce cost and provide better quality of work for the employees. As the internal cost goes down, the cost to your external customer can be reduced. As the quality

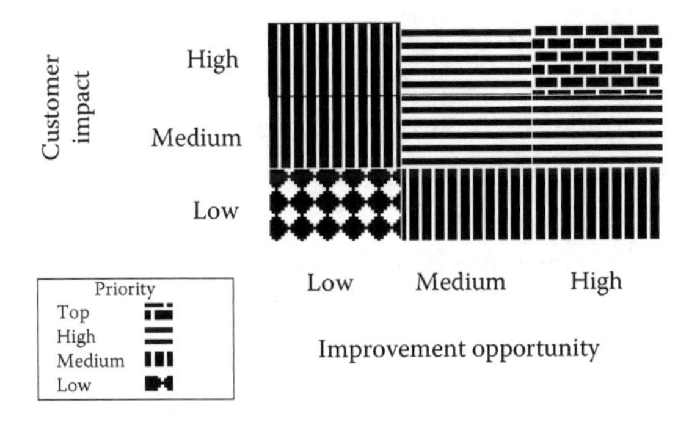

FIGURE 4.5
Process prioritization matrix.

of work life improves, the organization's output improves. Remember that every process, every activity, and every job within an organization exists for only one reason—to provide your customers and/or consumers with products and services that represent value to them. The rippling effect of improving any activity should have a positive impact upon the external customers.

Other typical factors that could be used are the following:

- Customer impact
- Sponsor support
- Changeability
- Quality
- Dollars saved

Weighted Selection Approach

Another way of accomplishing the same task is to have a group of executives give each of the LTM potential projects a rating of 1 to 5 in the following four categories:

1. Customer impact
2. Changeability
3. Opportunities
4. Business impact

A rating of 1 indicates that the process is too difficult to do anything with or it has little impact. A rating of 5 indicates that the process is very easy to change or it has a big impact. The rating of the four categories for each process and/or product should be totaled, and then these totals are used to set priorities, as shown in Table 4.1. This approach concentrates attention on critical issues, sets priorities for the resources, and assures that the effort is manageable. While it is a relatively simple and useful way to select LTM projects, this approach has a number of drawbacks, including the following:

- Pet projects are commonly identified.
- Management perspective may not be supported by hard facts.
- Top management may sway decisions.

A variation on the weighted selection approach is to weigh each of the individual categories based upon its impact on the organization, the customer, or the strategic plan. Typical weighting factors are the following:

- 1 = Low impact
- 3 = Medium impact
- 5 = High impact

The executive team would assign a weighting factor to each of the categories. The following is an example:

- Customers' impact = 5 weighting
- Changeability = 1 weighting
- Opportunities = 3 weighting
- Business impact = 3 weighting

TABLE 4.1

Weighted Selection Chart

Process Name	Changeability	Opportunities	Business Impact	Customer Impact	Total
Hiring process	3	4	5	4	16
Accounts payable	2	2	4	3	11
Engineering change release	5	3	5	3	16
Request for quotations	4	4	4	3	15
M5 product	5	5	5	4	19
New management training	4	3	3	3	13

TABLE 4.2

Weighted Selection Chart with the Process Rating and Category Weighting

Process Name	Changeability	Opportunities	Business Impact	Customer Impact	Total
Hiring process	1 × 3 = 3	3 × 4 = 12	3 × 5 = 15	5 × 4 = 20	50
Accounts payable	1 × 2 = 2	3 × 2 = 6	3 × 4 = 12	5 × 3 = 15	35
Engineering change release	1 × 5 = 5	3 × 3 = 9	3 × 5 = 15	5 × 3 = 15	44
Request for quotations	1 × 4 = 4	3 × 4 = 12	3 × 4 = 12	5 × 3 = 15	43
M5 product	1 × 5 = 5	3 × 5 = 15	3 × 5= 15	5 × 4 = 20	55
New management training	1 × 4 = 4	3 × 3 = 9	3 × 3 = 9	5 × 3 = 15	37

To get a value for any process/category, you multiply the category weighting factor by its rating. (For example, the value of the hiring process/business impact combination is business impact weighting 3 times the hiring process rating 5 = 15. (See Table 4.2.)

In the example shown in Table 4.2, the highest total score is the *M5 product* closely followed by the *hiring process*.

Another effective way of setting priorities is to complete a value analysis of the present activities and a second value analysis based upon the projected improvements and estimated cost to install and operate the change. A value analysis is calculated by dividing the cost to produce an output into the value of the output.

SUMMARY

As you complete your selection of the LTM projects, remember the four Rs.

- *Resources*—There are a limited amount of resources available, and the present processes must continue to operate while we are improving them. Often, this means that new processes or designs will be operated in parallel with old processes while new processes or new product designs are being verified. In many cases, it will be necessary to develop two levels of product at the same time. The evolutionary design will need to come out very quickly to prevent the erosion of the organization's market share while the revolutionary product is being developed and going through its analysis phases. Don't over-extend yourself.

- *Returns*—Look closely at the potential payback to the business. Will the process reduce cost? Will it make you more competitive? Will it give you a marketing advantage?
- *Risks*—Normally, the greater the change required, the greater the risks of failure. Major changes are always accompanied by resistance to change. Breakthrough activities have the biggest payback but have the biggest chance of failure.
- *Rewards*—What are the rewards for the employees and the LTM members working to improve the process? How much will their quality of life be improved? Will the assignment be challenging and provide them with growth opportunities?

The real key to a successful LTM program is the selection of the right improvement opportunities that can be handled by the LTM team. Watch out for the following when selecting an LTM project.

- Selecting projects that are too complex for the LTM
- Not selecting improvement opportunities that have the potential of creating real value added to the organization
- Not involving the key players who will make up the LTM team early in the process
- Management that will not agree to make decisions on the last day of the workshop
- Developing the list of potential LTM candidates without management knowing about it

Make the most out of every opportunity, and you will get the most out of life.

H. James Harrington

5

Phase II—Preparing for the Workshop

Prepare, not repair.

H. James Harrington

INTRODUCTION

After you have identified the improvement opportunities, you move into Phase II—Preparing for the workshop. (See Figure 5.1.) Preparing for the workshop is necessary because the project's sponsor needs to be identified and some additional work needs to be performed and data need to be collected in order for the workshop to be conducted effectively. Conducting the workshop should be done by a skilled facilitator. The time schedule for the workshop must be managed, or the workshop will not be completed on schedule.

There are four activities in Phase II:

1. Activity 1—Define the Lean TRIZ methodology (LTM) project sponsor
2. Activity 2—Hold a contracting meeting
3. Activity 3—Conduct a planning meeting
4. Activity 4—Collect relevant data to prepare for the workshop

Phase II is designed to accomplish five key tasks:

1. The first task is to make sure that the selected project will lend itself to the LTM project.
2. The second task is to get the LTM projects' sponsor identified, making sure he or she understands the LTM process and his or her agreement on what must be accomplished during the workshop.

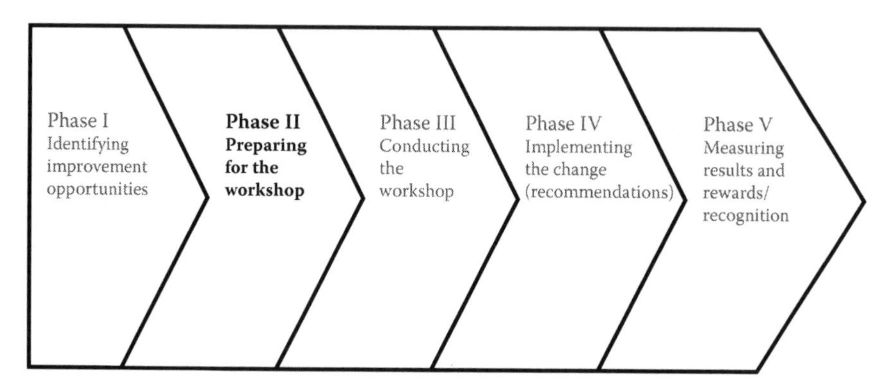

FIGURE 5.1
Phase II—Preparing for the workshop.

3. The third task is to define the workshop output expectations.
4. The fourth task is to identify the individuals who will make up the improvement team, gain their approval to work on the team, and gain their manager's approval to have them take part in the team activities.
5. The last task is to define the items' function and then to collect data that are adequate to define the cause of the problems.

The objectives of Phase II are the following:

- Confirm the viability of the LTM approach for the project
- Identify and gain support of the LTM project sponsor
- Attain sponsor's understanding of and commitment to the LTM project
- Develop the workshop scope and objectives
- Identify workshop functional expertise/participants' requirements
- Identify prework and other preparations necessary to conduct the workshop

Good things only happen when planned. Bad things happen on their own.

Phil Crosby
Quality Is Free, 1979

ACTIVITY 1: DEFINE THE LTM PROJECT SPONSOR

The LTM project sponsor is a manager who, due to his or her position in the organization, has the authority to authorize changes to the product, process, or subprocesses under investigation. The sponsor provides the resources to complete the LTM project activities and agrees to attend the last part of the workshop to listen to the recommended changes and make the final decision related to accepting or rejecting the suggested changes.

The LTM project sponsor is typically a middle-level manager who has a direct-line accountability relationship to the product or process that is being improved. Sometimes, due to the organizational structure, it may be necessary to have more than one sponsor for a single LTM project in order to provide the required resource commitments and decision on the recommendations. Often, the sponsor has already been selected during Phase I of the LTM process.

ACTIVITY 2: HOLD A CONTRACTING MEETING

The LTM champion should schedule a meeting with the potential sponsor to review the LTM project and to gain his or her concurrence to serve as the sponsor of the LTM project. The LTM champion should also determine if other people should attend the contracting meeting, i.e., potential change champions (team leaders), content experts, other impacted managers, etc. This is an important meeting because the sponsor needs to fully support the LTM project. For this reason, a well-prepared agenda is a must. The following is a typical agenda for an LTM contracting meeting (Figure 5.2).

To prepare for the contracting meeting, the LTM champion needs to collect information about the proposed product or process. He or she should have a basic understanding of the manufacturing process flow, some of the problems that the people are having, competitive products, and customer suggestions. The facilitator should have reviewed the individual performance indicators (IPIs) and understand how the product or process is measured and performing. The function of the output should be defined. This prework is necessary so that the facilitator and the LTM champion will sound knowledgeable when he or she meets with the sponsor. It will also greatly increase the effectiveness of the meeting and help to gain the confidence of the sponsor and the LTM team members in the LTM methodology.

The contracting meeting will start with the facilitator presenting an overall view of LTM to the sponsor and the other people attending the meeting. During the review, the facilitator needs to stress the expectations and responsibilities of the various roles in the LTM process with emphasis on the

Date: January 16, 2016

Time: 1:00 p.m. to 3:00 p.m.

Attendees:

Sponsor—Jack Johnson

LTM champion—Frank Voehl

Facilitator—Mary Thompson

Quality manager—M Robotic

Line product engineering manager—Abdul Awl

Agenda

1. LTM overview
2. Define potential LTM process and associated challenges
3. Evaluate product for application to the LTM approach
4. Develop improvement objectives
5. Define LTM roles
6. Define who should be involved in the decision-making process
7. Define primary planning team and team leaders
8. Develop a preliminary LTM workshop charter and process block diagram
9. Schedule the planning meeting and the workshop
10. Define the individuals who will participate in the workshop

FIGURE 5.2
Typical agenda for contracting meeting.

sponsor's role. The sponsor needs to understand that this process is participative and requires the commitment of the people in key roles beyond the sponsorship to ensure success. The facilitator will also need to set expectations as to the amount of detail and analysis that typically result from a workshop.

The facilitator should review the data that have already been collected. They will then ask the participants if they have any relevant data that can be contributed to help in taking advantage of the improvement opportunity. The facilitator will also ask, "What additional data are needed in order to define root causes?" The objective here is to have the facilitator and/or team leader collect sufficient data so that the LTM team will not require any additional information to define root causes and to make intelligent suggestions on how to take advantage of the improvement opportunity.

The starting point for defining and refining business process issues is the process flowchart. In most cases, during the contracting meeting, a very preliminary block diagram is prepared. This block diagram should be reviewed with the attendees to determine its validity and to add additional scope so that it can be transformed into a flowchart. (See Figure 5.3.)

Each step should be identified beginning with a verb. An effective way to do this is to have each of the attendees draw his or her own flowchart and then combine them together so that there is a common understanding. Try to limit the flowchart to no more than 10 major steps. Each of these steps should be recorded on a card that can be taped to the wall or to a blackboard. This will allow you to easily move them around and make connections. Key measurements related to each of the major steps should be added. This is just an estimate based upon the team's experience.

Process name: scheduling sales calls flowchart

Rev. 1

Inputs:

A. Phone call
B. E-mail

Outputs: make a sales call

Process activities:

1. Make a contact call to potential customer
2. Send out catalog
3. Make a follow-up phone call
4. Schedule a sales call

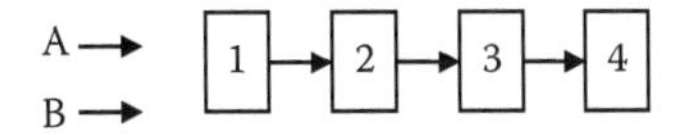

Issues:

- ❑ It costs too much to schedule a sales call.
- ❑ The scheduling cycle is too long.

FIGURE 5.3
Flowchart of scheduling sales calls.

Typical measurements would be cycle time, processing time, or cost. Now, the team can use the flowchart, looking at each step, to define issues and improvement opportunities. Have each attendee for each symbol on the flowchart record on an index card issues related to that symbol. A new card will be used for each symbol on the flowchart. These cards can be posted under the appropriate symbols so that a complete list is prepared for each symbol. Additional cards should be prepared for the process or subprocess as a whole and placed at the end of the flowchart.

Now, the team should go back and look at each step (symbol) to identify any existing process statistics that might be available and useful during the analysis. Typical statistics could be cost, cycle time, processing time, resources utilized, volumes processed, customer satisfaction, and quality performance. Individual attendees should be assigned the responsibility for going out and collecting these matrices so that they will be available for the workshop. As we work through this part of the meeting, it is important that the facilitator keeps the team members focused on problem identification, not problem-solving.

After a thorough discussion related to the LTM process, the discussion should shift to the targeted process or product for the workshop. The facilitator should lead the discussion for LTM process opportunities by preparing a high-level block diagram that shows the major activities that go on within the process. (See Figure 5.4 as an example of scheduling a sales call.) This high-level process block diagram should consist of no more than 5 to 10 blocks. In discussing the process, there are questions like

- Why was this process/product selected?
- What are the major issues and challenges related to this process/product?
- What will be the output from the workshop?
- What are the specific measurements that the workshop should focus on?
- Are there any parts of the process/product that are off limits and should not be evaluated?
- What are the targeted goals for improvement?
- What are the process boundaries?
- Why was the product selected?

If the LTM project is an evolutionary design analysis, the facilitator will lead a discussion related to the reasons why an evolutionary design

Process name: scheduling sales calls flowchart

Rev. 2

Inputs:

A. Phone call
B. E-mail
C. List from conferences

Outputs: make a sales call

Process activities:

Process activities	Cycle time	Process time
1. Sort by regions	24 hours	0.5 hours
2. Make a contact call to potential customer	85 hours	1.3 hours
3. Send out catalog	48 hours	0.2 hours
4. Follow-up phone call	120 hours	1.0 hours
5. Is a sales call justified?	1 hour	0.3 hour
6. Close out file	0 hour	0 hour
7. Schedule a sales call	24 hours	2.0 hours
8. Make sales call		
Total	302 hours	5.3 hours

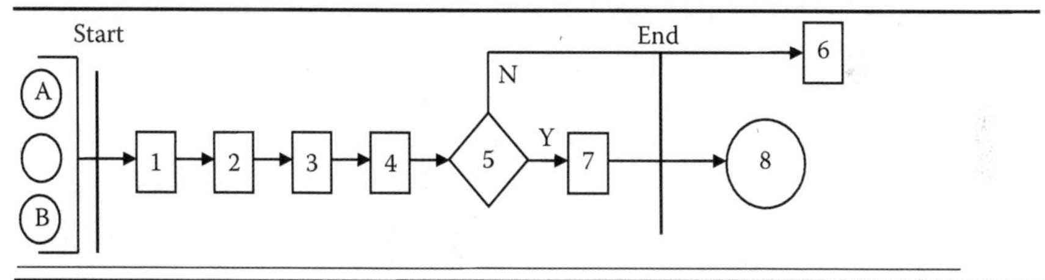

Issues:

- ❑ The scheduling cycle is too long.
- ❑ Too few sales per sales call.
- ❑ Sales people are not available.

FIGURE 5.4
High-level scheduling sales call block diagram.

change will be beneficial to the organization. Typical examples of things that would be discussed are the following:

- Sales trends
- Technology advancements
- Competitive activities
- Customer suggestions
- Low-profit margins
- Identification of new sales opportunities

Once there is common agreement about the process scope and boundaries, the group will discuss if the proposed process is a good candidate for an LTM-type analysis. In light of the sponsors' desired outcome, is the project scope within the capabilities of LTM, which is primarily achieving the objectives during the short, one- or two-day workshop, including analysis? If it's decided that the LTM workshop is not an appropriate approach, other alternatives should be identified to bring about the required changes. Product or processes that are outside the scope of the LTM program are ones that require heavy quantitative analysis, computer systems solutions, big investment in the corrective action, organization or major personnel changes, and things that are not in line with the organization's culture. During this meeting, the LTM champion obtains the sponsor's commitment as he or she will be very involved in the project. He or she will be responsible for the following:

- Providing the team leader to work with the facilitator.
- Kicking off the workshop.
- The sponsor attending the last part of the workshop where he or she will accept or reject the recommendations developed during the workshop. It is important that it is pointed out to the sponsor that he or she will need to make on-the-spot decisions based upon partially imprecise matrix.
- Determine if the sponsor is authorized to make the required decisions, or do we need additional sponsors to become part of the process?
- The sponsor will be responsible for attaining the resources required for the workshop.
- The sponsor will designate someone to track and oversee the implementation process, as well as communicate highlights to the appropriate management.

To put things in place and to document the direction of the workshop, a preliminary workshop charter will be prepared. (See Figure 5.5.) The preliminary charter should consist of five elements:

1. *Purpose*—This defines what the workshop is directed to accomplish.
2. *Targeted opportunity*—This defines the product or process that will be analyzed.

Sales call scheduling workshop charter rev. 1

Challenge:
To reduce the cost of scheduling a sales call
Area(s) of focus:
The process starts when a potential customer contacts us and ends when the sales call is scheduled.
Objectives:
To reduce the cost of scheduling a sales call
Team members:
Sales
IT
LTM sponsor(s):
Sales manager—Tom Jones

FIGURE 5.5
Preliminary sales call scheduling workshop charter.

3. *Objectives*—This defines the measurements that will be targeted for improvement along with improvement goals.
4. *Team members.*
5. *LTM sponsors.*

During this meeting the LTM champion and a sponsor will define when a planning meeting will be held and who will attend the planning meeting. They will also define who the team leader will be for the project. The team leader is an individual who is part of the process that will be working with the LTM members; he or she will be responsible for following up to be sure that the approved suggestions are implemented as scheduled.

Key Elements of an Effective Meeting

The following are some good rules to follow to have effective meetings:

- Adhere to a tight time schedule. Being late is no help to everyone who made it on time.
- The no-device rule should be enforced. No phones, virtual-reality headsets, laptops, or iPads should be allowed at the workshop in order to eliminate/minimize distractions. You can't be watching a screen or texting a message and be really involved in the team's activity. If you are distracted by these devices, it sends a message to the group that the team's activities are not important or interesting.
  - We have found that many people feel that this is a rule that they cannot adhere to, but the workshop should be designed to allow

five different opportunities for the participants to use these devices. They are the following:
  - Prior to the class starting.
  - During the morning break.
  - During the one-hour lunch break.
  - During the afternoon 15-minute break.
  - Immediately following the class.
- Provide a variety of recording material.
  - The LTM team should be provided with a variety of recording material to use to record their activities at the meeting. It's preferable to have too much rather than not enough flip charts and/or whiteboards. If your organization likes whiteboards, be sure that they are large and two sided. At a minimum, they should be 4 × 6 and on wheels. Also, be sure that they are cleaned prior to starting the class. We recommend that, for a typical LTM workshop, a minimum of one whiteboard per breakout group is recommended.
  - We personally like the old flip chart approach where each breakout group has their own flip chart holder. This allows the working groups to post a number of flip charts on the wall as they are working along and then pick them up and move them to the front of the room to show their output. When bigger sheets of paper are needed, butcher paper provides an acceptable alternative.
- Provide a projector and screen.
  - Often, the facilitator or the team leader will want to have a set of PowerPoint slides prepared prior to the meeting that will be used to acquaint the LTM team with the process they will be using. It's also often used as a time box monitor. Minute and second hands are projected on the screen with a special reminder just five minutes prior to the end of the time box. The computer projection setup is also often used for the presentation to the project sponsor. It is absolutely essential that the computer/projector combination that will be used during the class is checked out as being compatible and producing the desired results. Also, the lighting of the meeting room must be designed to ensure that the PowerPoint slides are easy to read while minimizing the drowsiness that the team would be subjected to if the lights over their working area were turned off.

- Provide general supplies.
    - Make sure that there are plenty of general office supplies available in the room. This includes things like the following:
        - Sticky notes
        - Colored markers
        - Filled-out name tags
        - Notebook paper
        - Masking tape
        - Large dot stickers
        - Felt-tip pens
        - Time clock
        - Coffee and snacks to keep the team's energy up
- Never take anything for granted.
    - The meeting facilitator should check out the room one hour prior to the start time of the LTM team workshop to make sure that everything is in place and working. We find it useful to provide the room setup crew with the picture showing exactly how he or she wants the room laid out. The meeting room should be completely laid out with name tags, pencils, markers, whiteboards, masking tape, etc., when the first person walks in to the workshop. All the equipment should be checked out ahead of time, and the correct PowerPoint slides should be loaded on the computer. Hot coffee should be located in the back of the room for those individuals who show up early. It's imperative that you have a well-organized facility setup ahead of the workshop to set the tone for the way the workshop will be managed and the way the team members will react to instructions.

In addition, the layout of the meeting room is extremely important to ensure the work moves smoothly without interruptions. Questions should be asked like, "Do we want roundtables, or should the room be set up like a classroom? Do we want flip charts or whiteboards?"

Also, it is important to make sure the participants are not allowed to be distracted. The facilitator should ask the question: Should personal cell phones and computers be allowed during the workshop? A study done by a group of 30 researchers from the University of California at Irvine indicated that each time an individual is distracted, it takes 23 minutes to get refocused and perform at the same level as before the distraction.

To keep the meeting on track, a regimented plan should be prepared. Otherwise, a two-day class might stretch into four or five days. The type of creative work that the LTM team will be doing requires long periods of uninterrupted blocks of time. If a meeting is supposed to start at 8 a.m. but doesn't start until 8:30, or if a 15-minute break is allowed to be extended to 30 minutes, the participants will feel like his or her time is not a precious commodity. Many organizations allow these delays to continually happen because it isn't in the organization's culture to be on a close time schedule. Living to a strict time schedule is one New Year's resolution we should all make and live up to.

Key Points Made during the Contracting Meeting

The following are some key points that should be embedded in everyone's mind as a result of the contracting meeting. They are the following:

- Limit the scope of the LTM project.
- Seek easy implementation solutions to the problems.
- Prepare sponsors to take immediate *yes-or-no* decisions at the end of the workshop.
- Explain lack of precise matrix, which decisions must be made on.
- Limit the size of the process being studied.
- Expensive system solutions, many information technology solutions, and reorganizational structure are out of the scope of the LTM initiative.

ACTIVITY 3: CONDUCT A PLANNING MEETING

We like to conduct this activity early in Phase II because the individuals attending the meeting often provide additional data and support. We recommend holding a preworkshop planning meeting for process-type problems. The purpose of the planning meeting is to add additional meat to the structure that was developed during the contract meeting. During the planning meeting, the following will be accomplished:

- The planning team is informed about the LTM process.
- The process block diagram is updated to a flowchart; it is validated and expanded.

- Key estimated performance measurements are added.
- Improvement opportunities/problems are identified.
- Preliminary charter is reviewed and updated.
- Workshop team is profiled, and participants are selected.
- The decision-maker list is finalized.

The following is a typical agenda for a planning meeting:

- Overview of the agenda and introductions.
- Overview of the LTM process.
- Review of the preliminary workshop charter.
- Review the reasons the specific product was targeted.
- Review and update the process flowcharts.
- Identification of improvement opportunities and problems.
- Review the charter scope and objectives to be sure they are still applicable.
- Quantify the objectives.
- Define the timing and structure for the workshop.

Holding the Planning Meeting

With the planning meeting scheduled, the sponsor has already approved the LTM project, and some general guidelines have been developed. The attendees at the planning meeting are usually more technical people as you can see from the agenda. These are people who are involved in the process or the product design, either in using it or in supporting changes and updates to it. There should be about 6–10 people who attend the planning meeting. The people who are already assigned to attend the LTM workshop should be in attendance. The average planning meeting lasts between one and two hours. It should start with a review of the agenda and the introduction of the attendees. Each attendee should state his or her background and how each is involved in the process.

The preliminary workshop charter that was developed during the contracting meeting is reviewed with the attendees. This provides them with an overall understanding of how a sponsor sees the problems and scope of the project. Also, a short presentation on the LTM process is provided so that everyone is starting from a common point of view and an understanding of what needs to be accomplished.

Typically, the types of people who will be involved in the meeting are looking for answers to problems, and, as a result, they have a tendency to want to solve problems. The facilitator should remind them that this team is focusing on identifying problems; solving the problems will be part of the workshop activities. If you get into problem-solving, the meeting will take far longer than it should. Ask them to share their suggested corrective action with the LTM team leader so that their ideas can be part of the action that comes out of the workshop.

The team should now go back and review the workshop objectives to see if they are complete and meaningful. In addition, target improvements should be added, i.e., reduce the total process cycle time from 300 hours to 200 hours or reduce process costs by 10%. In setting these goals, it is important to consider the intended scope of the workshop, realizing that it has to be accomplished in a maximum of two days. By the time this part of the meeting is over, there should be several quantifiable measurements that will gauge the success of the LTM workshop. Keep in mind that a typical workshop will make improvements on the order of 15% to 25%. It is unrealistic to look for bigger improvements that can be implemented in a 30-day period in most cases.

Next, the team will define the timing and strategy for the LTM workshop. In many cases, the workshop scope has expanded to the point that it may need to be broken up into a series of workshops rather than an individual workshop. Another consideration is, "Should the two days be consecutive or split, allowing the team to go out and collect some information?" We personally like to hold the LTM workshops off-site as it makes this special and gives the team an additional night to share ideas and to develop solutions. Often, we have held the workshops over the weekends so that people are not taken away from their normal assignments, but the off-site meeting is a nicety, not a necessity.

The team defines a preliminary date for the workshop and assigns the responsibilities for coordinating the logistics of the workshop to a team leader or a facilitator.

As we near the end of the planning meeting, the team has an excellent idea of what the LTM process is all about and the problems that need to be addressed during the workshop. With this information in hand, the planning team can profile what the workshop team should be like and identify the participants. In selecting the participants, the following should be considered:

- Are the individuals knowledgeable of the process/product?
- Will they be willing and able to participate in the workshop?

- Will they be willing to participate in implementing the suggestions that come out of the workshop?
- Are they creative by nature?
- Will there be someone there who can provide a view from the internal and external customers?
- Are all functions that are impacted by the process/product represented?
- Do you have the appropriate process–content expertise?

As a result of this discussion, a final list of individuals who will attend the workshop, along with their email addresses and phone numbers, should be compiled. Also, a plan should be developed on how they will be notified. Figure 5.6 is a sample workshop charter updated as a result of the planning meeting.

Once the workshop participants are identified, they should be instructed to seek input from their colleagues prior to the workshop. They will also be provided with an updated charter and a copy of the LTM book.

Sales call scheduling workshop charter rev. 2

Challenge:

To reduce the cost and cycle time of scheduling a sales call

Area(s) of focus:

Process inputs:

- ❑ Phone calls
- ❑ E-mail
- ❑ List from conferences

Process end: when the sales call is scheduled

Objectives:

- ❑ To reduce the cost of scheduling a sales call
- ❑ To reduce the cycle time from the first potential contact to the point when the sales call is scheduled by 25%

Team members:

Sales—Candy Rogers—Team leader
IT—Tom Jackson
Production control—Chuck Mignosa
Marketing—Johnny Black and Richard Jackson
Message center—Sue Brown
HME—Frank Voehl—Facilitator

LTM sponsor(s):

Sales manager—Tom Jones
Marketing manager—Brett Trusko

FIGURE 5.6
Updated workshop charter for sales call scheduling.

Situation questionnaire

1. Brief description of the challenge/situation objective

Describe your challenge/situation/objective in a single phrase.

2. Information about the system

2.1. System name

Name the system (product, process, technology, etc.).

2.2. System structure

Describe the system's structure. Be sure to indicate all important elements and how they are connected.

2.3. System functioning

Describe what the system was designed for and how it works.

2.4. System environment

Describe other systems that interact with your system.

3. Information about the problem situation

3.1. Problem that should be resolved

Describe the most challenging problem you are facing in pursuing your objective.

3.2. Factors causing the problem

Describe factors causing this problem.

3.3. Consequences of unresolved problem

Describe attempts to resolve the problem. State the reasons why these attempts were unsuccessful.

3.4. History of the problem

Describe attempts to resolve the problem. State the reasons why these attempts were unsuccessful.

4. Ideal vision of solution

Describe the ideal solution.

5. Allowable changes to the system

Describe any limitations for changing the system.

FIGURE 5.7
Typical simple situation analysis questions.

Filling Out the Situation Questionnaire

Very often in LTM, the planning meeting is eliminated by moving part of it to the contracting meeting and part of it to the workshop. This approach works extremely well when it comes to evolutionary design opportunities. Typical simple situation analysis questions were prepared by Ideation International, Inc. (See Figure 5.7.) They also use a more intensive situation questionnaire for more complex opportunities.

The situation analysis questionnaire will be filled out by the facilitator or the team leader well in advance of the LTM workshop. Copies of this analysis should be provided to all the members of the LTM workshop prior to the workshop. Completing the situation analysis will allow the team to quickly define the root cause of the problem and/or weaknesses in the design or process. To put it in TRIZ terms, filling out the situation analysis helps the team to find the major contradictions within a process and/or product.

Definition: Contradiction is defined as a situation that emerges when two opposing demands have to be met in order to provide the results required. A contradiction is a major obstacle to solving an inventive problem. It is often used as an abstract inventive problem model in a number of TRIZ tools. Three types of contradictions are currently used:

- Administrative
- Engineering
- Physical

Identifying contradictions and selecting approaches to offset the negative parts of the contradictions are a basic principle that the TRIZ methodology is based upon.

ACTIVITY 4: COLLECT RELEVANT DATA TO PREPARE FOR THE WORKSHOP

Careful preparation for the workshop is essential for completing the project within two days. The first thing that the facilitator should do is to prepare a letter that the sponsor will send out to the individuals assigned to the workshop explaining the importance of their attendance and participation. Attached to the letter should be a copy of the workshop charter. This letter should also instruct them to discuss the product/process being evaluated with their associates to define opportunities for improvement, problems, and potential fixes.

The next thing that the facilitator should do is to determine that the proper type of facilities is available. We like to set up the room with a u-type table arrangement with three roundtables in different corners of the room, which will be used as breakout areas. Each breakout area should have a flip chart, and there should be two flip charts in the front of the room. There should also be a PowerPoint projector and screen.

The next thing the facilitator should do is to prepare the agenda and review it with the team leader. When they are in agreement, it should be sent to all of the attendees, along with a technical report describing the LTM process or a copy of this book and a copy of the situation analysis. A typical two-day agenda is as follows:

- Kick-off and introduction.
- Business challenges defined by the sponsor.
- Expectations for the workshop defined by the sponsor.
- LTM overview.
- Workshop charter.
- Review the situation questionnaire results and the data that have been collected related to the item's function and other key measurements and then agree on root causes and/or major constraints.
- Redefine process map or redefine as-is matrix if it is applicable.
- Identification of issues, opportunities, and problems.
- Prioritization of issues, opportunities, and problems.
- Define and quantify actions to address issues, opportunities, and problems.
- Prioritize/classify action items.
- Perform a cost–benefit analysis.
- Prepare the sponsor's presentation.
- Present to the sponsor.
- Summarize results and document them.

The agenda should take into consideration any special cultural requirements (e.g., time to pray, don't work on Sundays). It is usually good practice to meet with the sponsor to jointly develop his or her setting expectations talk and preparing a set of PowerPoint slides for him or her to use.

On the day before the workshop, the facilitator should contact everyone to be sure that they are coming to the workshop and that they know the right time and place of the workshop and to find out if they have any questions or special requirements (e.g., the individual is a vegetarian, the individual cannot climb steps).

By now, the stage is set for the workshop. The LTM champion and facilitator have done everything that they can do to set up the props and prepare the actors.

Either the project team leader or the LTM facilitator needs to have a background in the TRIZ methodology. Because there is no formal degree program from any major university related to the TRIZ methodology, the TRIZ professional organization, Altshuller Institute, has established a four-level certification program that is used to classify the skill level of TRIZ practitioners. The following are the skill levels listed, from the lowest (1) to the highest (4) skill level:

1. Associate
2. Practitioner
3. Specialist
4. TRIZ masters

At a very minimum, the skill level of either or both the LTM facilitator and the project team leader should be at the TRIZ practitioner level. Although some of the individuals certified at the associate level are very competent to do the job, there are others who have little or no experience in completing a TRIZ process.

Now, it is time to pull up the curtain and put the creativity of the team members into action.

SUMMARY

Phase II is extremely important as it sets the stage for the workshop. First of all, LTM methodology is reviewed, and the sponsor agrees to personally be involved with the implementation of the agreed-upon improvement opportunities.

The second critical activity that takes place during Phase II is collecting pertinent data that allow the root cause of the opportunity to be identified.

By the end of this phase, the facilitator and the team leader should both agree on whether the assignment is a process improvement opportunity or a product redesign opportunity. If these two items are not done in a very professional manner, the LTM process is doomed to fail.

> A good guide is better than any map.
>
> **H. James Harrington**

6

Phase III—Conducting the Workshop

Time-box the day, and you will get the most out of it.

H. James Harrington

After you have identified the improvement opportunities, you move into Phase III—Conducting the Workshop. (See Figure 6.1.) This is the phase where creativity and natural abilities need to kick in.

There are six activities used to define the future-state solution, which can be further divided into 13 tasks.

- *Phase I: Identifying improvement opportunities*

  It is absolutely crucial that the project sponsor understand the role he or she must play and the quantity and type of resources that will be needed for the project.
  - Activity 1: Introduction
    - *Task 1*: Introduction and business overview given by the sponsor
    - *Task 2*: Present an overview of Lean TRIZ methodology (LTM)
- *Phase II: Preparing for the workshop*
  - Activity 2: Develop the as-is model
    - *Task 3*: Reviewing the situation questionnaire results and agree on root causes and major contradictions related to the improvement opportunity
    - *Task 4*: Refining the process flowchart or reviewing the current product specifications
      - *Task 4.A*: Refining the process flowchart
      - *Task 4.B*: Reviewing the current product specifications

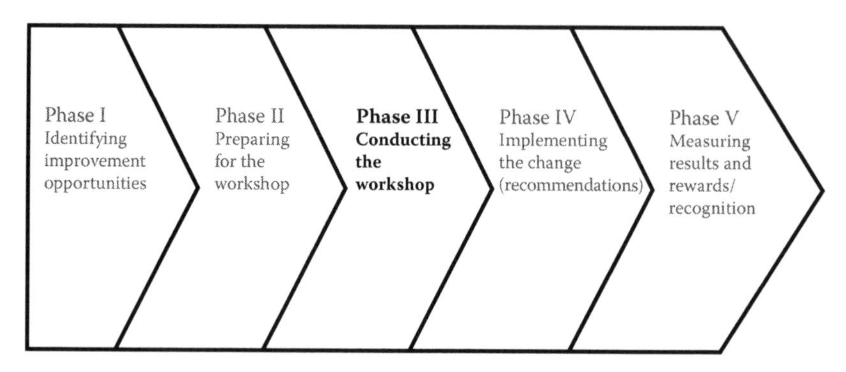

FIGURE 6.1
Phase III—Conducting the workshop.

- *Task 5*: Refining the as-is matrix
    - *Task 5.A*: Refining the process as-is matrix
    - *Task 5.B*: Refining the product as-is matrix

- *Phase III: Conducting the workshop*
    - Activity 3: Start workshop and define root causes
        - *Task 6*: Identifying issues, opportunities, problems, and root causes that were not covered in Task 5
            - *Task 6.A*: Identifying process issues, opportunities, problems, root causes, and major contradictions
            - *Task 6.B*: Identifying product issues, opportunities, problems, root causes, and major contradictions
        - *Task 7*: Prioritizing issues, opportunities, problems, and root causes
    - Activity 4: Define potential improvement options
        - *Task 8*: Defining and quantifying improvement action for issues, opportunities, problems, root causes, and major contradictions
            - *Task 8.A*: For *process design/redesign improvement opportunities*, use processing tools like brainstorming, streamlined process improvement (SPI), Lean, and Lean Process TRIZ (LP-TRIZ)
            - *Task 8.B*: For *product design/redesign improvement opportunities*, use brainstorming, value engineering, and TRIZ
    - Activity 5: Quantify improvement opportunities
        - *Task 9*: Prioritizing and classifying improvement action items
        - *Task 10*: Quantifying and qualifying potential improvement solutions

- Activity 6: Management review and approval
  - *Task 11*: Preparing the sponsor's presentation
  - *Task 12*: Presenting to the sponsor
  - *Task 13*: Summarizing the results and documenting them

It is important to remember that there are two alternatives to the way the workshop can be conducted. They are the following:

1. Process design/redesign improvement opportunities
2. Product design/redesign improvement opportunities

Both of these approaches use a five-phase LTM approach. Phases I, II, IV, and V are essentially the same in either approach. Phase III—conducting the workshop—is where there is a real difference between improving the process and improving the product. (See Figure 6.2.)

The purpose of the workshop is to develop the action plans to improve the targeted process/product that can be implemented during the next 30 days. The workshop starts with the sponsor providing a big-picture overview of the key challenges that face the business, particularly as they relate to the desired changes and how LTM contributes to the organization's goals. From this point, the workshop focuses on understanding the process or product, its problems, and developing process and/or design improvements that the workshop attendees can implement in the next 30 days. Once the improvement opportunities are defined, their benefits are quantified and presented to the sponsor

Product approach	Process approach
• Activity 1. Introduction	• Activity 1. Introduction
• Activity 2. Develop as-is model	• Activity 2. Develop as-is model
• Activity 3. Start workshop and define root causes	• Activity 3. Start workshop and define root causes
• Activity 4. Define potential improvement solutions using brainstorming, value engineering, and TRIZ	• Activity 4. Define potential improvement solutions using brainstorming, streamlined process improvement, Lean, and LP-TRIZ
• Activity 5. Quantify improvement opportunities	• Activity 5. Quantify improvement opportunities
• Activity 6. Management review and approval	• Activity 6. Management review and approval

FIGURE 6.2
Comparison of the phases for process and product design.

for approval. All of this happens within a two-day period. To accomplish this, minimum time is spent on developing objectives and charters. The two days are time-boxed. Sufficient data should have been collected that allow a group to quickly agree on the root cause of the opportunity.

Definition: Timeboxing is a way to manage an individual or group's time. Timeboxing allocates a fixed time period, called a *time box*, to each planned activity. Several project management approaches use timeboxing. It is also used for individual use to address personal tasks in a smaller time frame. It often involves having deliverables and deadlines, which will improve the productivity of the user.

Now, you can see that there is a lot to be accomplished in two days. It will not happen unless the facilitator manages the time very effectively. To do this, the facilitator will use a technique called *timeboxing*. Figure 6.3 shows the percent of accomplishments that are made during a typical 1-hour meeting that is not time-boxed.

Looking at Figure 6.3 depicting accomplishments in a typical meeting, you will note that usually the meetings start slow and just before the end of the meeting must end, things start to happen very fast. However, when you time-box a meeting, there is a significant increase in accomplishments

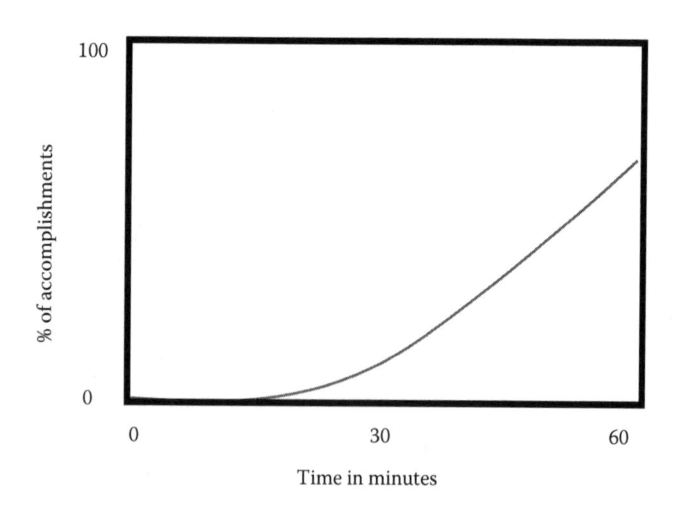

FIGURE 6.3
Accomplishments versus time of meeting without timeboxing.

in a shorter period of time as shown in Figure 6.4. In Figure 6.4, the team was able to accomplish as much in 30 minutes as it was able to accomplish in 60 minutes before they used timeboxing. In both cases, they didn't accomplish everything that they would like to accomplish. Due to time restrictions, the meeting is usually ended with still some open items that could be discussed.

Timeboxing is a technique where things are divided into small time periods that are strictly adhered to. The workshop starts on time, and breaks are closely controlled and scheduled. Timeboxes are used as a form of time management to explicitly identify uncertain task/time relationships, i.e., the individual activities that make up the workshop activities' work may easily extend past its deadline. Time constraints are often a primary driver in completing the workshop in one to two days and should not be changed without considering project or subproject critical paths. We like to use a large 1-hour timing clock set up in a position where everyone on the team can readily see how much time they have to complete the specific assignment. The clock is set to give a 2-minute warning prior to the end of the specific time box. Time is the only resource that can't be replaced once you have used it.

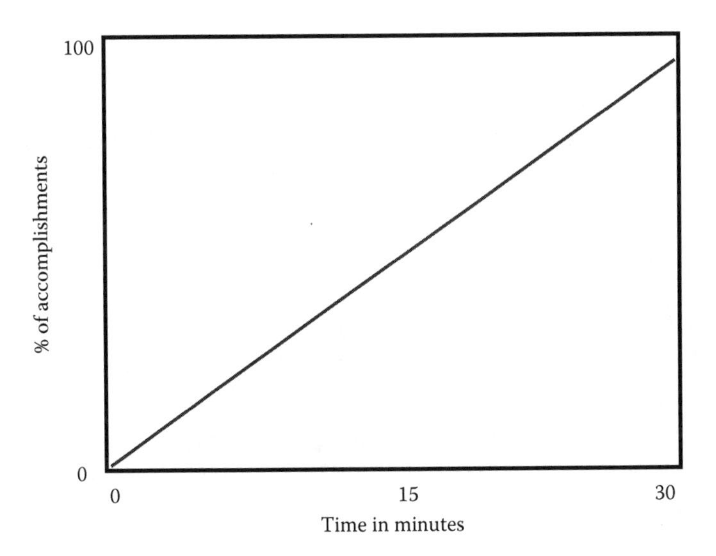

FIGURE 6.4
Accomplishments versus time of meeting with timeboxing.

PHASE I: IDENTIFYING IMPROVEMENT OPPORTUNITIES

Activity 1: Introduction

Task 1: Introduction and Business Overview

The team leader should review the agenda and have each attendee introduce himself or herself and state the organization that they work for and explain how they are involved in the process. The sponsor will then present a business overview that will outline the business case, including the scope, goals, and objectives of the workshop. If possible, he or she should tie this in with the strategic plan for the organization and the competition the organization is facing. This is the sponsor's opportunity to update the participants in the relevant business challenges. The sponsor will also review the workshop charter, discussing specific goals, objectives, and why the team was selected and list the key performance measurements. Following this, the participants will be given a chance to ask questions and then the sponsor will leave the workshop.

Task 2: Present an Overview of LTM

Task 2 is designed to provide the team with an understanding of what they will be involved in for the next two days. Particular emphasis will be placed on finding solutions that they can implement over a 30-day period. They will also focus on the timeboxing approach and the need to have a proposed future-state solution available to present to management on the last day of the workshop.

Probably the most effective way to gain a good understanding related to a process and/or the way a product is processed is through the use of a control chart and performing a process walk-through.

Collect Pertinent Data

This activity is typically handled by the project facilitator and/or the team leader. It is a critical and time-consuming activity as it provides a database that the workshop decisions will be based upon. The data that are collected vary based upon the type of improvement that the LTM team is working on.

Typical data that might be collected in a product design or redesign activity are the following:

- Recent technology changes
- Sales volume for the last six months

- Competition's product performance data
- S-curve analysis
- Customer complaints
- List of suggested product improvements
- Results of focus group with potential customers
- Product-return rates
- Build cycle time

Typical data that might be collected for a process improvement or redesign activity are the following:

- Process flow diagram
- Process cycle time
- Customer complaints
- Cost to process and item
- Workflow between work areas
- Transportation cost and cycle time

PHASE II: PREPARING FOR THE WORKSHOP

Activity 2: Develop the As-Is Model

Task 3: Review Situation Questionnaire Results and Agree on Root Causes and/or Major Contradictions Related to the Improvement Opportunity

Definition: Situation questionnaire is a document used to collect information related to the improvement opportunity. It is usually a set of questions that the organization feels is necessary to define the root cause of a problem or opportunity. It has a dual purpose:

- It provides a checklist of the information necessary to define a root cause
- It provides a way to collect data related to the proposed questions.

One of the big advantages of LTM is the detail that goes into collecting information related to the improvement opportunity before the workshop starts. The situation questionnaire is designed keeping in mind the old saying, "An improvement opportunity well-defined is half-solved." In LTM, the project facilitator or the project manager has the responsibility of investigating the situation related to the improvement opportunity. This includes collecting relevant data that shows the magnitude of the opportunity and the potential root causes that make the opportunity available. He or she should be particularly diligent in defining what the major contradictions are related to the opportunity that the team will be working on. To accomplish this, the good as well as the bad part of the improvement opportunity must be defined. Sometimes a positive part of the improvement opportunity creates a contradiction that can be very harmful. Doing a good job in establishing the as-is conditions and contradictions related to the improvement opportunity is critical to the success of the LTM project. A poorly researched situation analysis can cause the LTM team to solve a problem that is nonexistent and/or to come up with a partial, temporary solution or a solution to a problem that does not exist.

Definition: A *contradiction* occurs when two opposing demands have to be met in order to provide the desired results. A contradiction results in a major obstacle being identified that must be addressed to solve an inventive opportunity. It uses an abstract inventive problem model and a number of TRIZ tools. There are three types of contradictions that are used in the TRIZ model: (1) administrative, (2) engineering, and (3) physical (as defined in the TRIZ methodology).

For product-related opportunities, the team should review the related engineering drawings and manufacturing routings. A product quality–related opportunity can result from a change in the engineering drawings and/or specifications or from a change in the way the product is manufactured. The facilitator should have the defective part inspected to ensure its tolerances fall within the allowed engineering specifications. Often, what looks like a product problem is, in reality, a manufacturing process problem. When this is not the case, the ideal situation is for the facilitator to be able to simulate the problem, allowing the engineering team to model potential corrections. We have observed problems created under a variety of situations—when a new operator was assigned to an operation and he or she was having trouble with the equipment, when an operator decided that the manufacturing setup instructions did not need to be and were thus not followed, and when an operator was afraid of telling his or her manager that he or she made some parts that were not to specification, and when the manufacturing routings and/or setup instructions are not in sync with the latest engineering change levels, etc. For all of these reasons and more, when we're having a product problem, we need to understand how the product is produced. This often involves preparing a flow diagram of the related manufacturing processes and then doing a walk-through to ensure that the manufacturing routings and setup instructions are being followed and in keeping with the latest engineering release level. In evolutionary designs conversion time is a major factor. One of the best ways to get an evolutionary design out to your customer is by studying the old products' production cycle and using it to create the new product. Setting up a new production line is costly and time consuming.

If it is a cost or productivity improvement opportunity, a manufacturing flow diagram is also very helpful in identifying potential no-value-added (NVA) activities. This activity in the cycle should not be treated lightly. The LTM team should feel confident that they have defined the real root cause of the opportunity and/or the major contradictions that need to be offset in order to have an optimum solution.

Task 4: Refining the Process Flowchart or Reviewing the Current Product Specifications

Task 4.A: Refining the Process Flowchart

The purpose of this task is to complete the flow diagram which will serve as a primary tool in process improvement and to prepare a complete list of the relevant measurements based upon the current products' prints and specifications.

Before the workshop, the facilitator should take each of the flowchart activities and record the information on a 5 × 8 index card. (In some cases, a 4 × 6 card will be adequate.) The first step in this activity is to get the team to agree on the start and end points of the assigned process. The modern business organization comprises a complex maze of interactive, interconnected, and/or sequential processes. Defining the process boundaries to break the maze into logical management pieces is a critical task. In most processes, the beginning and end points are not clearly defined. One person might view them in a limited sense and another in a more global manner. As a result, it is important that the workshop team agree on the beginning and end points. Consider, for example, a familiar activity like barbequeing a steak. Based upon the process statement, there are many points at which the process could begin. Some of the beginning points could be the following:

- Getting a job so that you have money to buy the steak
- Deciding to have a steak barbeque
- Going to the store to buy the food
- Taking the steak out of the refrigerator to cook it

The process also has many potential end points. They could be one of the following:

- Process is finished when the steak is cooked.
- Process is finished when the steak is served.
- Process is finished when the steak is consumed.
- Process is finished when the mess is cleaned up and put away. (I got this one from my wife Marguerite.)

As you can see, there are many potential combinations of beginning and end points. To start the discussion, the facilitator can point out the beginning and end boundaries that the workshop charter was based upon. When the team agrees on the start and end points, the facilitator should put up the first process flowchart activity index card on a whiteboard or the wall. The workshop team should discuss this activity and agree on what the first process activity should be and what inputs trigger that activity. The facilitator leads the workshop team through the process flowchart, putting up new index cards as the flowchart develops and changing the preliminary process flow to reflect the team's more detailed understanding of the process. The facilitator should not allow the process flowchart to become too detailed. Particular attention should be paid to troublesome and/or high-cost activities.

For product redesign, the team will evaluate both the engineering and production processes that relate to the product.

Task 4.B: Review the Current Product Specifications

The team will make a list of all relevant specifications and requirements. This is normally done by reviewing the current specifications and prints to identify engineering requirements. Particular emphasis will be placed on those requirements that directly impact the external customer. The

LTM team is usually assigned to improve one or a combination of the four parameters. They are the following:

1. *Competitive analysis*—Competitive analysis is developing a new product that is competitive on the market. In this case, the facilitator will develop a list of the key parameters related to the product being improved. He or she will also often collect actual performance data related to the competition.
2. *Cost*—Reduce the cost of the product to make it more competitive and to increase profits for the organization. To satisfy this objective, the techniques specified in value engineering provide a valuable tool for identifying potential cost-saving.
3. *Quality*—Eliminate a specific problem that is often impacting cost and schedule. Attacking quality problems presents a bigger challenge than cost reduction as the improvement activity may be related to process and/or the product. As a result, one or more of the performance improvement tools may be used. Typically, quality-related improvements are qualified by collecting customer complaint information and/or defect rates.
4. *Productivity*—Increase the value added (VA) per employee. Improving productivity from a product standpoint is driven by concepts defined in designed-for manufacturability. To accomplish this, it is valuable to make use of a process flowchart that has the processing time for each activity recorded on it. Once this has been accomplished, the LTM team will focus its efforts on how the design can be changed to make it easier and less complicated to produce. Often the final design is modified so that the product or part of the product can be produced using robotics in place of human effort.

Task 5: Refining the As-Is Matrix

Task 5.A: Refining the Process As-Is Matrix

The purpose of Task 5 is to provide quantitative data related to the flowcharts that were prepared in Task 4. These data will be used to define potential improvement approaches when analyzing targeted processes. In preparing for the workshop, the facilitator and/or the team leader should have collected the majority of the data by analyzing the elements in the process flowchart like costs, processing time, cycle time throughput capabilities, and reject rates. It is often necessary to validate these data by conducting a process walk-through by the facilitator and the team leader where they validate the information that they have received. Doing this walk-through provides an excellent opportunity for the team leader and facilitator to spend time with the person actually doing the work and get ideas from them on how the process can be improved. This is the point in the cycle where an activity flowchart or knowledge flowchart can provide significant additional input to the LTM team.

Task 5.B: Refining the Product As-Is Matrix

For product improvements, the team will define and quantify functional parameters related to product improvement. For these efforts, the facilitator and/or the team leader should have reviewed the prints and specifications to identify key functional parameters. With the product improvement cycle, it is important that good information is collected from the current customers and potential future customers. Closely examining the collected data often reveals gaps in the data-collection process. These gaps need to be filled for the project to be successful. In addition to the as-is matrix, the LTM team needs to use its crystal ball and project how the technology will change during the cycle time when the product is being developed, documented, manufactured, and ready for delivery to the customer. For the project to be successful, the LTM team needs to design a product that is better than your competitor's product to increase your market share and gain customer loyalty.

PHASE III: CONDUCTING THE WORKSHOP

Activity 3: Start Workshop and Define Root Causes

Task 6: Identifying Issues, Opportunities, Problems, Root Causes, and Major Contradictions that Were Not Covered in Task 5

Task 6.A: Identifying Process Issues, Opportunities, Problems, Root Causes, and Major Contradictions

During this task, the LTM team will be using the four methodologies that are effective at developing a new process or redesigning the present process. The four methodologies are (1) brainstorming, (2) SPI, (3) Lean, and (4) LP-TRIZ.

Using the flowchart for the targeted process constructed during Task 4, the team will develop the relevant data. For each block and connecting line between each block, the team should estimate the value of the key parameters. For each block, the team needs to make the best estimate on what the key matrix related to that block is. Typical matrices could be the cycle time, quality, items per hour, etc.

The LTM team may have already provided some inputs, but these inputs should now be refined. The team should not worry about getting exact figures since the best estimate that the team can provide is usually adequate. They will also identify the key part of the process that limits the process output capabilities. These throughput limiting operations are excellent points to start any improvement action.

- *Brainstorming*

  The facilitator will start by giving the team green dot cards and leading the team through a brainstorming session to define the improvement actions. These improvement actions are recorded on index cards, computers using a projector, flip charts, or whiteboards. The same approach is then applied to the yellow- and orange-colored index cards until the allotted time is used up. This is a very time-consuming task, and it will require everyone to be creative and original thinkers. Although the facilitator will encourage the team to define improvement actions that are easy to implement by the team members, ideas that are outside of the scope of the LTM project should be recorded as they provide good input to the management team. The biggest problem the facilitator faces during this activity is to keep the team saying, "*This is what we can do!*" rather than, "*This is what they should do.*" Be sure not to rule out a good idea just because it had failed before. Sometimes, conditions have changed, and things that did not work before will work now.

  Now that there is an agreement related to the process flowchart, the facilitator will start with the first block of the flowchart and ask this question: "Are there any issues, improvements, problems, root causes, or major contradictions that were not previously defined related to this project? If there are, record each of them on a separate index card. Classify each card as an issue, an improvement opportunity, a problem, a root cause, or a contradiction. Be sure that you aren't making any improvement recommendations. The team leader should fill out cards for the items pointed out by the six task activities. When the cards have been completed, the facilitator has each attendee read one card. The facilitator should then place the card in relevant groupings. This process will be repeated until all the cards have been posted. The LTM team, with the aid of the facilitator, will combine any relevant groupings that overlap each other.

After developing the appropriate flowchart(s) related to the process improvement opportunities and quantifying the key functional parameters related to a process improvement project, the LTM team should be able to define the critical problems through brainstorming and through the use of the Five Why's approach. The LTM team is now prepared to start developing improvement approaches related to the key performance issues.

- *SPI*

  How to use SPI, Lean, and LP-TRIZ were all covered in detail in Chapter 2 of this book. As a result, we will not be providing detailed information about any of these three improvement approaches in this chapter.

  As we start to use the SPI methodology, we should have already created a fairly accurate flow diagram of the process with costs, cycle time, processing time, and defect rate information related to each activity. This flow diagram and data package plays a critical role in optimizing the process. Major process improvement changes will be largely based upon this information. The LTM team should start this activity by reviewing the flowchart and the associated data. Once this is completed, they will start using the 12 improvement approaches defined in the SPI approach. They are the following:

  1. *Bureaucracy elimination*—We like to start with bureaucracy elimination for this is something that everyone would like to see eliminated. As you review each activity on the flowchart, you should color in the bureaucracy with a blue highlighter. We then look at each of the blue activities to determine if there is any VA as a result of the activity. Often, the cost of the activity far outweighs the value to the organization.
  2. *Value-added analysis*—In using this tool, we focus on classifying each activity into one of the three following classifications:
     1. Real value added (RVA)
     2. Business value added (BVA)
     3. NVA

     The objective of this activity is to eliminate the NVA activities to minimize the BVA activities and maximize the RVA. Many of the processes within most organizations have no RVA activities.

3. *Duplication elimination*—The objective of this analysis is to eliminate duplication wherever it is possible. This includes recording of the same information on different forms and keeping similar data in two or more places. For example, the manufacturing manager keeps a record of the number of units he or she shipped. Production control keeps a similar database. All too often, due to timing or other reasons, the numbers that go to the executive team are different. This causes unnecessary delays and expenses as the differences are accounted for.
4. *Simplification*—All too often, we try to use technology to replace an activity that is being performed by a human being when it is not only not needed but also often more costly. In other cases, we use the latest and best technology instead of making use with what we have on hand. This not only causes process delays but also increases the costs to perform the activity. For example, look at what happened to a breakfast cereal–producing organization that was getting complaints back from its customers that the box it ships was only half full. Being customer focused, they immediately went out and bought an x-ray machine and a robot. The x-ray machine made every box, and, if the box wasn't completely full, the robot was programmed to remove the box from the conveyor line. This solved the problem, and everyone was happy. At another location, while the manufacturing engineers were waiting for delivery of their x-ray equipment and the new robot, a lady on the production line brought a fan from home and set it up so that it would blow the half-filled boxes off the production line and into a salvage bin.
5. *Cycle time analysis*—The organization worries about and focuses on processing time because that is money directly related to the cost of the product. Cycle time is the time the customer waits for their order to be fulfilled. If the organization is working to adjust the time schedule, this is often four times longer than the processing time. The LTM team should look at any time that the paperwork and/ or the product is not moving forward. Typical delays are stocking, transportation, and equipment bottlenecks.
6. *Error prevention*—Often called *negative analysis*, as the LTM team looks at the process to define what could be done to cause the process to fail. When these failure modes are defined, the

team evaluates them to determine how the process needs to be modified to prevent them from occurring.

7. *Supplier partnerships*—We know of no organization that doesn't depend upon suppliers in one way or another. The supplier–customer relationship is really a partnership, and each individual organization needs to go 65% of the way. Don't automatically assume that a purchased item that doesn't work is a supplier problem. As much as 35% of these are caused by the customer rather than the supplier. In dealing with suppliers, consider how you would feel walking in their shoes and what you would want. They want long contracts, maximum cycle time requirements, and a minimum of engineering changes. Schedule your parts from the supplier so that they come in when you need them, not before where you pay for storage, and not too late where you slow down production.
8. *Technology*—Some organizations in developing countries can survive without some level of technology built into their processes. It doesn't have to be the latest and best. In fact, sometimes, upgrading to the latest change globally is NVA. Every time you look at the possibility of buying technology, do a thorough value analysis. Consider not only the costs of buying and installing the technology but also the impact upon your personnel as they learn to use the new technology. We talk about technology including things like robotics, automation, and mechanization. Consider how fast the technology is changing; in many cases, the whole life cycle of the technology is two years or less.

   When you're offered software packages like customer relationship management (CRM), the salesperson indicates that it will provide you with a significant competitive advantage that will greatly improve your market share. Don't depend upon this because, probably, your competition is also installing a CRM system, giving you no competitive advantage. And it will cost the organization a great deal of money and time to purchase, install, and train its people. Should you do it? Yes! If you don't do it, and your competition does it, it will put you at a very serious disadvantage.
9. *Process upgrading*—It is often the little things that make a big difference. Consider the following: How bright is the work area? What is the color of the department's walls? How far is it to the

coffee machine? How much traffic is there from the test to the work area? How much and what kind of filling space do you have? Is your work area as pleasant as the work areas in other departments? What kind of music is being played? How comfortable is the chair that you are sitting in? Remember that the working environment is an individual's home away from home; in reality, most people spend more time in their work environment than they spend in their home environment.

The author offers the following example. He can remember an incident with one of his inspectors (Employee A) who found a powerful magnifier that was mounted to a stand. As he made his daily visit to each employee, he noticed that Employee A was using different equipment, and he asked her why. She explained that she could do a better job, and it was easier on her eyes. He congratulated her on being very innovative, and, as soon as he got back to his office, he ordered one for each of the two other inspectors. A week later, the other two magnifiers came in, and it looked like everybody was happy until he noticed that Employee A was not looking at him when she said good morning. After a week of this behavior, he brought her into his office to ask her what is wrong. He began by explaining that she had a brilliant idea that was so good that he purchased the same equipment for the other inspectors. He was surprised to find out that Employee A was unhappy because the other two ladies had new magnifiers and she was using an old one. She felt that it was her idea, and, if anyone was to get a new magnifier, it should've been her.

PS—She got a new one in two days.

10. *Risk management*—By this point in the cycle, the team should've come up with a number of improvement ideas. Some of these ideas are evolutionary, and some of them are revolutionary. Now, the team steps back and reviews each improvement suggestion to determine what risks are related to the specific activity. Each new idea should be evaluated as high risk, medium risk, or no risk. High-risk items should be evaluated to determine if a medium- or low-risk approach would have the same impact upon the process.

    Often, the biggest risks that an organization faces is the resistance employees have to change. Resisting change is a normal and natural thing for everyone since very few people will embrace personal change. Almost everyone is enthusiastic about other people changing but don't feel he or she needs to change. Putting in an effective change management system is absolutely essential to offset this negative tendency to fight and resist change.

11. *Standardization*—It is very difficult to make progress in improving a process if everyone isn't doing it the same way. Evaluate the improvement ideas to determine if everyone can follow the same instructions. If not, modify your approaches so that everyone can do the same thing in the same way. You cannot improve if everyone is doing the job in different ways!
12. *Simple language (improve communication)*—Communication is one of the biggest problems that most organizations face. It's difficult enough if you're communicating with people who have the same background and education level. Unfortunately, most of the process specifications are written by college graduates and are used by people where English is a second language and may only have a high school diploma. All documents should be written at the 10th grade level, making them 20% faster to read and have a 25% better retention level.

- *Lean*

  The Lean methodology is directed at reducing waste. Brainstorming and streamlining the improvement process are also both directed at the elimination of wastes. As a result, many of the tools that make up the Lean methodology have already been covered in the brainstorming and streamlining activities. Since the tools that make up the basic Lean package are documented in Chapter 2, the LTM team should look at these tools to determine if any of them need to be applied to the process being studied. Typical new tools are introduced in one-minute change of dies, just in time, equipment maintenance procedures, the cleanliness of the work area, and single-unit flow production lines.
- *LP-TRIZ*

  LP-TRIZ is a methodology that focuses on three major targets—(1) productivity, (2) cost, and (3) quality. It uses a 3 × 41 matrix to evaluate where a particular change initiative will have the greatest impact (productivity, cost, or quality). The 41 elements in the matrix are the 41 actions that are most often used to bring about performance improvement in a process. Each one is analyzed for its impact on productivity cost and schedule. Often, a specific individual activity can have a positive impact on one of those three targets and a negative impact on another. When this occurs, the team analyzes the matrix to identify an activity that has a positive impact upon the negative activity without having a negative impact on the previous positive activity. Using this approach, the team will evaluate each of the proposed improvement ideas to determine its total impact upon the system. For more information, see Chapter 2.

Task 6.B: Identifying Product Issues, Opportunities, Problems, Root Causes, and Major Contradictions

In a process improvement program, the emphasis centers around the flowchart as it provides a road map that leads from the as-is state to the future state for the process. For product improvement programs, the flow diagram plays a much less important role. The chief emphasis in a product improvement program is the functional parameters and how they need to change to be competitive. In the LTM, when we talk about functional parameters, we are not talking only about data that are defined in prints and/or specifications. What we are talking about is the parameters that the customer buys the product for. This frequently requires the collection of data that are not specified in the engineering product release documentation.

Another key indicator is an analysis that determines where the present product fits on the S-curve. Products that are at the top of the S-curve or sliding down on the right-hand side of the S-curve are excellent candidates for Lean TRIZ. The last important data source is all the data collected through the interfaces with the customer (sales, delivery, cycle time, nonfunctional quality conditions). Another excellent source of information related to new product development is a comparison of the current product versus its competitors. Typically, both marketing and development engineering will have already gathered this information. It is the reason that you should have a marketing representative on the LTM team when you are developing new products or evolving already-established products.

The data that are collected vary based upon the type of improvement that the LTM team is working on. Typical data that might be collected in a product design activity are the following:

- Recent technology changes
- Sales volume for the past six months
- Competition product performance data
- S-curve analysis
- Customer complaints
- List of those suggested product improvements

- Results of focus group with potential customers
- Product return rates
- Build cycle time

If the LTM team is working on an evolutionary product design improvement, the as-is matrix would be very different from the process matrix. For example, the as-is matrix would look more like the following:

M Robotics line pickup improvement as-is analysis.

- Round unit has a print of all the item pickup capabilities—3.00 to 1.00 inches
- Square item pickup capabilities—5.00 to 1.00 inches
- Speed—25 units per minute*
- Error rate—3 units per 1,000 units processed
- Claw material—rubber grade 407*
- Positioning approach—vibration
- Product size—58 × 48 inches
- Arm reach—63 inches
- Weight—68 pounds*
- Power consumption—250 W*
- Average percent uptime—98%*
- Controller technology—SMT3
- Average set up time—22 minutes*
- Costs to build—$8321
- Maximum claw pressure—35 torque pounds*
- Maximum pickup weight—10 pounds*
- Placement location accuracy—plus or minus 0.005 inches*
- Production schedule for the next three years
- Set of prints and schematics for the unit

In list above, the items marked with an asterisk are considered functional items as they have a direct impact upon the external customer. For new product improvement, the most valuable information comes from discussions and surveys with a customer, consumers' product performance, and projected technology changes.

Task 7: Prioritizing the Issues, Opportunities, Problems, and Root Causes

If the team has identified many issues, improvement opportunities, problems, and/or root causes, it is often difficult to address all of them in the time available during the workshop. As a result, it is useful to prioritize these items. To do this, a two-dimensional matrix can be developed with the left-hand side of the matrix entitled "Impact upon the Process" or "Impact on the Organization." It is divided into three segments—(1) high, (2) medium, and (3) low. (See Figure 6.5.)

Each of the nine blocks in Figure 6.5 has a number recorded on it. Blocks 2, 3, and 6 are the high-priority potential solutions and should be addressed by the LTM team. Blocks 1, 5, and 9 are the second-order opportunity, and the team should select the ones that would be addressed. The bottom side of the matrix (Blocks 4, 7, and 8) are the low-priority solutions and should not be addressed by most teams.

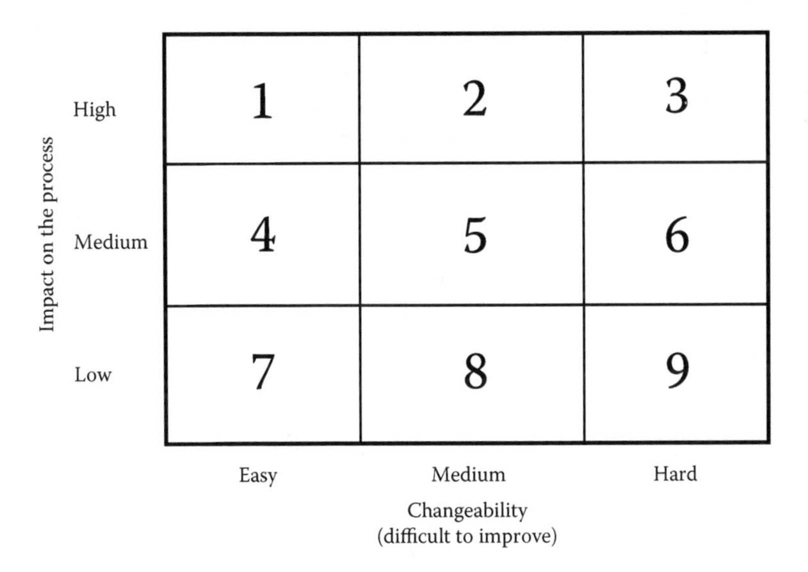

FIGURE 6.5
Prioritization of issues, improvement opportunities, problems, root causes, and contradictions.

Sometimes, when there are a lot of issues, opportunities, problems, and root causes, a number system is used instead of words. Along the horizontal axis, when we are rating the improvement solution—instability—as easy, medium, or hard, we use nine numbers on the scale. In this case, *easy* is replaced with 1, 4, and 7; *medium* is replaced with 8, 5, and 2; and *hard* is replaced with 9, 6, and 3. Along the vertical scale (impact on the process/product), low impact is replaced by 7, 8, and 9. Medium impact is replaced by 4, 5, and 6, and high impact is replaced by 1, 2, and 3. This allows the LTM team to refine their estimates and weight a potential improvement from 1 to 81. This technique is usually used when there are a lot of very good potential improvement solutions.

In addition to the input from the situation questionnaire, we often use the 5Ws approach to transform issues, improvement opportunities, and problems into root causes in Tasks 5 and 6. This activity often surprises the LTM team when they find how many of the issues, improvement opportunities, and problems relate back to a single root cause. In using this approach, the team asks why the condition exists by asking *why* at least five times or until the root cause has been identified. Then, they ask themselves what impact correcting this would have on the process and/or product. This approach should give the team enough understanding to place each root cause in one of the nine boxes. We like to place a paper dot on the index card indicating its priority level.

- A green dot equals *highest priority* (1s).
- A yellow dot equals *high priority* (2s).
- An orange dot equals *medium priority* (3s).
- A black dot equals *low priority* (4s).
- A red dot equals don't do (5s) unless there is an excellent reason to include it.

This nine-box approach is an excellent starting point, but don't let it override your collective knowledge of the item being improved. We like to complete the nine-box application and define what the high- and medium-priority root causes are. Then, ask the LTM team if there are any other improvement options that didn't make the improvement list that should be included. We have seen LTM teams that blindly follow the nine-block method to define the root causes that will have improvement actions prepared by them. Employee gut feeling is sometimes better than any more formal way to identify and prioritize the improvement opportunities.

PHASE III: CONDUCTING THE WORKSHOP

Activity 4: Define Potential Improvement Options

Task 8: Defining and Quantifying Improvement Action for Root Causes

The LTM team has to be very creative in coming up with solutions and should not be constrained by the current culture, personalities, or environment. The team should keep in mind the following points as they focus on qualifying and selecting the potential improvement solutions that will be presented to the sponsor:

- Rework can be eliminated only by removing the causes of the errors.
- Moving documents and information can be minimized by combining operations, moving people closer together, or automation.
- Waiting time can be minimized by combining operations, balancing workloads, or automation.
- Expediting and troubleshooting can be reduced only by identifying and eliminating the root causes.
- NVA outputs can be eliminated if management agrees.
- Reviews and approval can be eliminated by changes in policies and procedures.
- Reliability problems can be eliminated by selecting better suppliers, more efficient heatsink, designing parallel circuits, using different material, etc.

Challenge everything. There is no sacred cow in LTM. Every activity can always be done in a better way. The end result of this process analysis is an increase in the proportion of RVA activities, a decrease in the proportion of BVA activities, a minimizing of NVA activities, and a greatly reduced cycle time. From a product standpoint, it can result in a design that skips one development cycle, allowing the organization to come out with products that are significantly better than the competition.

This is a very time-consuming task, and it will require everyone to be creative and original thinkers. Although the facilitator will encourage the team to define improvement actions that are easy to implement by the team members, ideas that are outside of the scope of the LTM project should be recorded as they provide good input to the management team. The biggest problem the facilitator faces during this activity is to keep the team saying,

"*This is what we can do!*" rather than, "*This is what they should do.*" Be sure not to rule out a good idea just because it had failed before. Sometimes, conditions have changed, and things that did not work before will work now.

Task 8.A: For Process Design/Redesign Improvement Opportunities, Use Brainstorming, SPI, Lean, and LP-TRIZ

The facilitator should point out to the team that they should look for time wasters like reports, approvals, meetings, measures, politics, practices, etc. They should then think about each of the items to see if they could be eliminated, partially eliminated, delegated downward, done less often, done in a less complicated way, less time-consuming manner, or done with fewer people. If the card approach is used, each card should have the name of the person who filled out the card on it. The cards will be placed under the root causes they relate to. In most cases, there are two slightly different approaches used—one approach for defining improvement activities for processes and a slightly different approach for defining improvement activities for evolutionary designs.

For process design/redesign improvement opportunities, we use four unique tools to create a process' future-state solution.

1. Brainstorming
2. SPI methodology
3. Lean methodology
4. LTM

These improvement methodologies were discussed earlier in this book, so we will provide only key thought patterns related to the methodology to refresh your memory. If you need more details, read Chapter 2.

- *Brainstorming*

  We like to start the problem-solving part of the workshop using brainstorming as a tool to get everybody involved and feeling comfortable with speaking up when they have an idea. All too often, the team's activities are centered on ideas generated by one or two people with the rest of the team going along with it. In addition, some good basic thoughts and ideas are generated during a short brainstorming session. They are the following:
  - *Activity 1*—Collect and share as-is status information.
  - *Activity 2*—Define root causes for the improvement opportunity.

- *Activity 3*—Generate potential ideas that will correct the as-is problem that can lead to the creation of a best-value solution state.
- *Activity 4*—Prioritize potential solution.
- *Activity 5*—Discuss advantages and disadvantages for each potential solution.
- *Activity 6*—Perform a risk analysis (advantages and disadvantages of the proposed solutions).
- *Activity 7*—Develop an action plan with people assigned to implement each of the activities in the action plan.
- *Activity 8*—Set up a measurement system and implement the proposed changes.

- *SPI*

Streamlined process improvement (often called business process improvement) is a well-proven concept for improving process performance. For the LTM, only one of its five phases of methodology will be used—Phase III: Streamlining the processes. The SPI methodology usually provides the LTM team with a number of improvement options. The following lists shows the 12 tasks that make up one of the activities in Phase III—Streamlining the processes.

1. Bureaucracy elimination
2. Value-added analysis
3. Duplication elimination
4. Simplification
5. Cycle time reduction
6. Error proofing
7. Supplier partnerships
8. Technology
9. Process upgrading
10. Risk management
11. Standardization
12. Readability (simple language)

One of the most effective tools in streamlining the process is value process analysis. It focuses upon classifying each step in the flowchart in one of three categories. They are the following:

- VA
- BVA
- NVA

- *Lean*

  The Lean approach was discussed in Chapter 2 of this book. It focuses primarily on manufacturing processes, but many of the things can be applied across all processes (e.g., picking up and putting away things that are not used, having a place for most used things to save time in finding them). Chapter 2 provides the readers with a checklist of tools that can be applied. Some of them are not covered in streamlined improvement. One word of caution—don't select any improvement technique that does not add real value to the organization. For example, installing a clean desk philosophy where every desk has all the papers picked up and put away prior to going home. We've seen organizations where this 5S approach has had a negative impact upon the value of the operations, although it does improve the look within the departments.
- *LTM*

  The LTM is the derivation of the TRIZ 39 × 39 Contradiction Matrix. It uses a 41 × 3 matrix with 41 frequently used approaches to improving processes. (See Figure 6.6.)

  In the LP-TRIZ approach, you define which one of these three parameters (quality, cost, and productivity) are the most important to improve as you move horizontally across the matrix. The junction point will provide you with an estimate if that particular change to the process will have a positive, medium, or no impact upon the desired results. LP-TRIZ is covered in detail in Chapter 2 of this book; the writer suggests that you go back and review it at this point in time.

Task 8.B: For Product Design/Redesign Improvement Opportunities, Use Brainstorming, Value Engineering, and TRIZ

During this task, the LTM team will be effective if they use the three methodologies that are effective at developing new products or redesigning the present product. The three methodologies are brainstorming, value engineering, and TRIZ.

- *Brainstorming*

  The facilitator or the team leader will instruct the LTM team to take three minutes to record two ideas of how they believe the process could be improved. The facilitator will then record each idea

1	Eliminate operations
2	Combine operations
3	Use smaller lot sizes
4	Do away with stock
5	Eliminate all bureaucracy
6	Redo the time standards
7	Change the requirements
8	Use first in, first out
9	Perform a better risk analysis
10	Start using a sampling plan
11	Track surpluses
12	Relay out the work area
13	*Fast* dept. reorganization
14	Pick up
15	Improve sales force
16	Train people
17	Relocate people
18	Recognition
19	Prepare job descriptions
20	Put in an employee certification system
21	Get rid of layers of management
22	Work with suppliers
23	Get new suppliers
24	Supply chain management
25	Customer understanding
26	ISO 9000
27	Empowerment
28	Provide better parking for customers
29	Independent evaluation
30	Rearrange tools
31	Get rid of things not used
32	Add more equipment
33	Computerize the process
34	Better maintenance
35	Belts to help move more parts
36	Make new tools and dyes
37	Visual tracking
38	Simple language
39	Computerization
40	Online stock levels
41	Innovation

FIGURE 6.6
The 41 most frequently used ways to improve processes.

on a whiteboard or a flip chart as each individual shares his or her idea. If two or more people have the same idea, the facilitator will just indicate the number of times the idea had been expressed without recording it again. The class will be provided with sticky dots of three different colors—red, yellow, and green. The team will be instructed to place the green dot behind the idea that they thought was the best. They then place a yellow dot behind their second choice and the red dot behind their third choice. The facilitator will then calculate a point score for each idea by crediting each green dot with three points, each yellow dot with two points, and each red dot with one point. By summing together the point score for each idea, the facilitator will come up with an individual point score for each idea. The three to five ideas with the highest-point scores will be carried on to the next part of this analysis.

- *Value engineering*

  Value engineering goes hand in hand with value analysis. Value engineering provides users with a list of best practices related to product type. Value analysis is the logical way to measure value-added productivity.

  For the sake of this example, let's assume that there are nine members of the LTM team. In this case, we would divide the team members up into three groups of three people each. The facilitator or the team leader would have previously divided the appropriate lists of best practices into three separate groups. Each of the three-person teams would be responsible for analyzing one-third of the value engineering best-practices list. Each team will analyze each of the assigned items to determine if it would have a positive effect on the product and present their findings to the rest of the LTM team. The facilitator or the team leader will record each team's conclusions on a flip chart or a whiteboard. The team will then analyze the list to determine if any of the items on the list should be combined. Each LTM member will be provided with three sticky dots that they will post by the action items that they feel are the most important. The green dot represents his or her first choice, yellow dot—second choice, and red dot—third choice. To weigh each of the action items, the facilitator will calculate a weighted number giving three points for a green dot, two points for a yellow, and one point for red dots. The team will then discuss the findings and double-check that they did not miss any important action item. The team can add good ideas to the list of potential corrective actions as they come up during the workshop.

- *TRIZ*

  The LTM team does not have time to use the more sophisticated TRIZ approaches, but, for our purposes, just working with the Contradiction Matrix greatly improves the final products. TRIZ is covered in more detail in Chapter 2. The following is a quick summary explaining how TRIZ is used in the LTM. It is going to be primarily used to highlight problems related to the proposed corrective action approaches.

  The TRIZ Contradiction Matrix is a 39 × 39 matrix. The vertical column is the characteristic that you are trying to improve. The horizontal column is labeled the same as the vertical column, but, in this case, it is the characteristic that will get worse as a result of the proposed improvement effort. The following is an example of how the LTM team would use the Contradiction Matrix.

  The team decides that they will decrease the weight of the wheel frame to increase the car's speed. Weight is the first item on the left-hand side of the vertical column. As a result, it is number one of the 39 characteristics that are improving. As we scan horizontally across the matrix, we find junction points where the proposed corrective action could conflict with a characteristic, thus making that characteristic perform at a lower level. In this case, when changing the weight of a mobile object, there are potentially 14 contradictions in the first 21 (21/39) characteristics listed horizontally on the page. (See Table 6.1.)

  - The team defines which of the 39 characteristics (top of the 39 × 3 Matrix) that could improve the item's performance.
  - The team analyzes each of the contradiction points listed across the horizontal axis and classifies them as high probability of occurring, medium probability of occurring, and low probability of occurring.
  - For each of the high probability of occurring, they analyze the present principles that are listed at the junction point.
  - The principle that is most likely to minimize the impact of the contradiction is then added to the list of potential improvements.
  - This cycle is repeated until all the high-risk items have been addressed. In some cases, the team will also look at select medium probability of occurring conditions the same as they did the high probability of occurring conditions.

TABLE 6.1

Characteristic That Is Improving

Characteristic That Is Improving	TRIZ 40 Principles
3. Length of a mobile object	15, 8, 29, 34
5. Area of a mobile object	29, 17, 38, 34
7. Volume of a mobile object	29, 2, 40, 28
9. Speed	2, 8, 15, 38
10. Force	8, 10, 18, 37
11. Tension/pressure	10, 36, 37, 40
12. Shape	10, 14, 35, 40
13. Stability of composition	1, 35, 19, 39
14. Strength	28, 27, 18, 40
15. Time of action of a moving object	5, 34, 31, 35
17. Temperature	6, 29, 4, 38
18. Brightness	19, 1, 32
19. Energy spent by a moving object	35, 12, 34, 31
21. Power	12, 36, 18, 31

We have spent our time seeking the best improvement ideas from the three different sources quality, cost, and productivity. In each of these sources, we have selected the low-hanging fruit to make a better apple pie. These different perspectives should have created some new and robust approaches to improving the situation. In a normal improvement situation, the team spends 20% of their time generating one solution and then spends 80% of their time justifying their decision. To accomplish this, we had to quickly harvest the fruit from each of the three trees when there is another fruit that is ripe and could be picked. The small-group approach allows individual team members to express his or her opinion without tying up the conversation into a long debate where the individuals are trying to defend their position.

As a result of this multifaceted approach, the LTM team will come up with a variety of approaches to solving this problem—some of them short term and others long term. As a result, the team should evaluate each of the proposed improvement concepts and classify them as one of the following two classifications:

1. Short-term improvement solution
2. Long-term improvement solution

The long-term improvement concepts should be put on the parking board so that they can be presented to the sponsor for further consideration.

Now, with this list of high-potential, short-term improvement solutions, the team is ready to narrow down the list to the most promising solutions. Of all the tasks in LTM, this is the one that is the most difficult to live up to the time box plan.

PHASE III: CONDUCTING THE WORKSHOP

Activity 5: Quantify Improvement Opportunities

Task 9: Prioritizing and Classifying Improvement Action Items

The result of Activities 6 and 7 is a list of improvement suggestions, some of which may be clearly outside the scope of the LTM project. Others are good ideas but may not have the return on investment to justify the implementation. To focus the team on the best alternatives, each of the improvement suggestions could be prioritized using a two-level matrix. One index is entitled *Implementation Effort*, and the other is *Associated Returns*. (See Figure 6.7.)

Using the prioritization and classification matrix, we will again use the dot technique to show the classification.

- *Green dot*—Bell ringers (easy to implement/big payback).
- *Yellow dot*—Quick hits (easy to implement/small payback).
- *Orange dot*—Process with care (tough to implement/big payback). These may be outside of the LTM project scope.
- *Red dot*—Stay away from these (tough to implement/small payback).

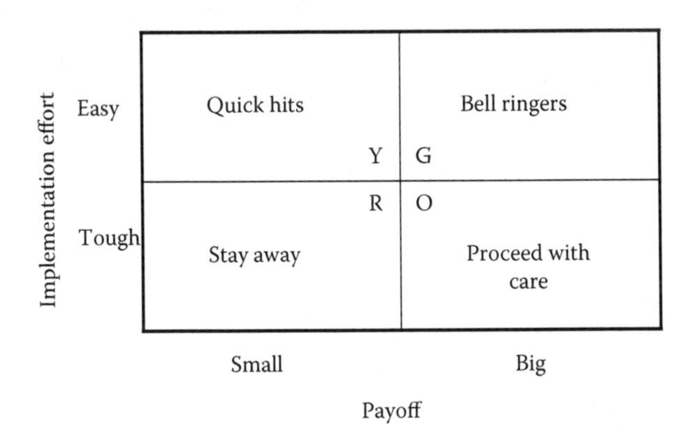

FIGURE 6.7
Prioritization and classification matrix.

This has now turned the whiteboard or the wall into a rainbow of color that can be quickly used to guide you to the most important and effective solution.

At a number of places in this book, we presented different prioritization methods, many of which were more accurate and precise than the one we just discussed. In many cases, the additional accuracy was justified because the data that were used were more precise. The combination of the prioritization method with the team's gut feel is adequate for this point in the improvement process.

Task 10: Quantifying and Qualifying Potential Improvement Solutions

In Task 9, the LTM team selected potential improvement solutions that would provide the greatest improvement in the item they were assigned to improve. But that isn't enough to decide if the solution should be implemented. Improving performances is only one piece in the puzzle. There are at least three other parts to the puzzle to solve in order to make a decision to implement or not implement a specific solution. They are the following:

- Cost to implement the solution
- Cycle time to implement the solution
- Acceptability of the solution

Cost to Implement the Solution

To accomplish this, the LTM team will prepare a block diagram flowchart using their combined knowledge of what it would take to implement the solution. Support activities like meetings and training should also be included in the block diagram. Don't try to get too much detail in the box diagram. Typically, 6 to 14 boxes will make up the diagram. In most cases, just a time-oriented list is adequate. The following is a typical example of the headings in a box diagram for the release of the evolutionary design to a current product:

- Product engineering builds a mock-up of the new product.
- Product test evaluates the performance of the new products and estimates its reliability.
- Marketing develops a marketing plan for the new product.
- Paperwork is prepared and released from product engineering to manufacturing.

- Projected manufacturing cost estimates were prepared by production control.
- A new product is announced.
- Suppliers are selected, and model parts are inspected and received.
- Manufacturing engineering prepares routings, job instructions, and manufacturing layouts.
- A production model is constructed and tested to the engineering specifications.
- Production control releases work orders to manufacturing and to purchasing.
- Manufacturing people are trained.
- Marketing and sales implement their sales campaign.
- The LTM team holds weekly two-hour meetings.

The author suggests that a person from finance be invited to be on the team to help make cost estimates as the next step is to estimate the number of employee hours required to do each of the items in the block diagram. The team then multiplies the controllable costs in each area by the number of hours estimated to be spent on this project. In addition to this cost analysis, any special equipment, services, software, etc., that are required to do the job should be estimated and included.

Of course this is the ideal situation and in the time box limits LTM team will not be able to construct box diagrams for every potential improvement solution. This approach is only used on high-cost mission-critical solutions. For the less important ones, an estimate from knowledgeable people in the team is adequate. We are looking for estimates that are accurate to plus or minus 25%. One word of caution—the facilitator should be careful to ensure that all of their cost estimates are on the low side and all of their savings estimates are on the high side.

Cycle Time to Implement the Solution

The cycle time to implement the solution is also an important factor. We would like to see all of the approved solutions implemented in 30–90 days. To get an estimate of the implementation cycle time, the team will place the block diagram headings on the left-hand side of the time-oriented spreadsheet. For each activity on the spreadsheet, there will be a cycle time recorded taking into consideration any imports that need to be provided before the activity can be completed. Figure 6.8 is a typical 60-day plan.

Action = ▬
Ongoing activity = ////////

Act. #	Activity/person(s) responsible	Week ending dates													Person(s) responsible
P	Planning	4/5	4/12	4/19	4/26	5/3	5/10	5/17	5/24	5/31	6/7	6/14	6/21	6/28	
1.0	3-year/90-day plan														
1.1	Transcribe EIT written plan														E&Y/KL
1.2	Draft 90-day plan														E&Y/KL
1.3	Mail draft to EIT and division czars														E&Y/KL
1.4	EIT modifies/approves plan														EIT/NCH
1.5	Return revised plan to E and Y														EIT/NCH
1.6	Revise plan as needed														E&Y/KL
1.7	Present final plan to EIT			X		A	I								E&Y/KL
2.0	Develop individual 3-year plans														E&Y/Div. czar
ML	Management support/leadership														
1.0	Teams														
1.1	Establish task team														EIT/NCH
1.1.1	Identify needed training														Task team
1.1.2	Develop training plan					A									Task team
1.1.3	Establish training budget						R								Task team
1.1.4	Present budget and plan to EIT							A							Task team
1.3	Implement training plan								I						Div. pres.
3.0	Strategic direction														NCH
4.0	Performance plan and appraisal														
4.1	New appraisal processes								R						RAH
6.0	Suggestion system														
6.1	Establish task team														EIT/NCH
6.2	Determine type of suggestion system														Task team

X = Review F = Form R = Report I = Implement P = Plan T = Train A = Approval D = Done

FIGURE 6.8
Typical 60-day plan for an evolutionary new product.

You will note at the left-hand side of the spreadsheet is the name of the individual and/or organization responsible for that assignment. It is sometimes necessary to move a potential improvement activity from the recommended list to the parking board when the cycle is too long.

Acceptability of the Solution

Improvement efforts that do not have the support of the affected personnel will not be as effective as they should be. Unfortunately, often, improvements that are very effective never get implemented. Everyone is for change. He should change. She should change. The company should change. But don't ask me to change—I've been successful doing just what I've been doing for the past five years. Why should I take a chance on something new if I'm already doing a good job?

Resistance to change is a very natural reaction. It provides the organization with the benefit of flavor-of-the-month–type programs because it requires the support of these programs to question its real value.

> Only 5% of the organizations in the West truly excel. Their secret is not what they do but how they do it. They are the ones that manage the change process.
>
> **H. James Harrington**

Most of your employees are probably already overwhelmed with the increased acceleration of change in their professional and private lives. Therefore, managers must realize that organizational change can and must be managed. Change cannot be viewed as a one-time event or a passing phase. Change must be seen as the manageable process which it is. (See Figure 6.9.)

The LTM team must be capable of guiding the affected individuals safely through the change process that involves convincing people to leave their comfort of the current state, move through a turbulent, new way of doing things (the transient state), to arrive at what may be an unclear distant future state. Specifically, these three states are defined as follows:

1. *Current state*—The status quo, or the established pattern of expectations. The normal routine an organization follows before an improvement is implemented.
2. *Transition state*—The point in the change process where people break away from the status quo. They no longer behave as they have done in

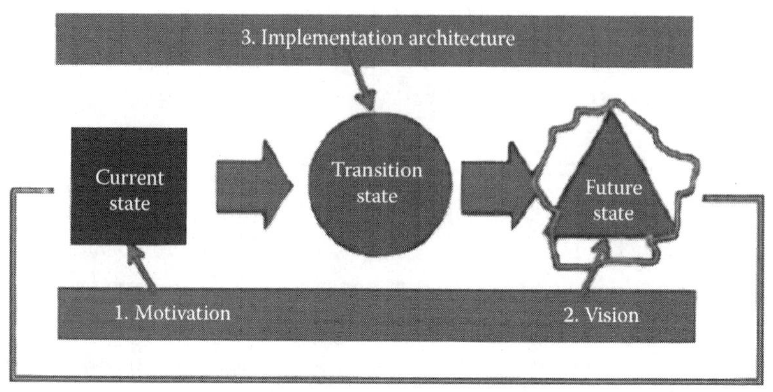

FIGURE 6.9
The process of change.

the past, yet they still have not thoroughly established the *new way* of operating. The transition state begins when solutions disturb the individuals' expectation, and they must start to change the way they work.

3. *Future state*—The point where change initiatives are implemented and integrated with the behavioral patterns that are required by the change. Team goals and objectives have been achieved.

We like to think of change as a teeter-totter. (See Figure 6.10.)

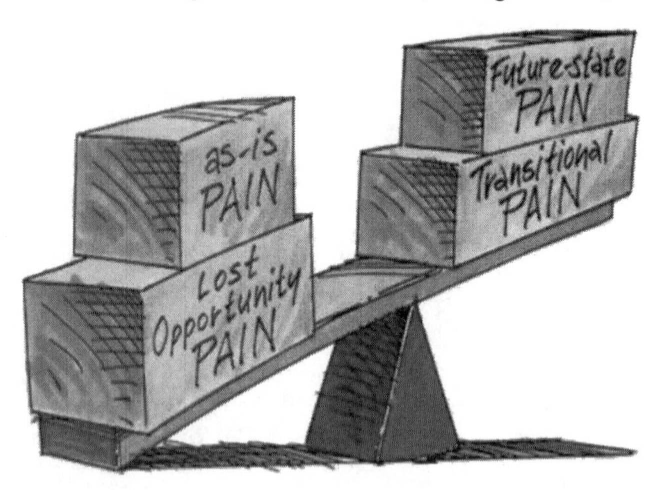

FIGURE 6.10
Organizational change teeter-totter.

On the one end of a teeter-totter is the pain related to the job the individual is currently performing. The pain is understood by the employee probably better than anyone else as he or she lives with it. But this is only part of the current pain. We also have to consider what would happen if the improvement change was not implemented and we kept on doing what we're doing in the same old way. This could result in the loss of market share and jobs within the organization. It's up to the sponsor and the LTM team to make the individuals aware of what pain they could be subjected to if they don't change.

At the other end of the teeter-totter is the pain that they will be subjected to when the change is implemented. It is the LTM team and manager's responsibility to help the employees understand the magnitude of the future state pain. There also is pain related to the transitional period when the changes are being implemented and the employees are beginning to learn how to use it. For the employees to be supportive of an improvement solution, they need to feel that there is more pain related to the current state than there will be as a result of the transition state plus the future state. To make this happen, there has to be a very clear view of what the future-state conditions will be. Without this vision, it is like an individual driving 70 miles an hour down a winding road with the windshield covered with mud. (See Figure 6.11.)

Step I

Define what will change.

FIGURE 6.11
A driver with a clear vision.

The Reality of Estimating the Value of Improvement

The previous three subsections provided a very theoretical view of how to determine the value of a proposed improvement activity. In reality, few of the LTM teams have the time and/or the need to do these types of evaluations for most of the improvement initiatives. Instead, the team will need to divide the improvement opportunity into a series of small opportunities that can be implemented rapidly. For instance,

- Change the routing so that assembly goes to Department K prior to going to Department W.
- Change all the screws to American standard.
- Replace a machined gear with a molded plastic gear.
- Rearrange office layout to improve workflow.
- Add a unit assembled at the suppliers.
- Eliminate the job report as it is already on the computer.
- Eliminate equipment purchase owners' sign off by the controller for budget equipment.
- Put in tool boards throughout the manufacturing area.
- Go to skip lot inspection for receiving inspection.
- Change the marketing strategy to focus on future products rather than current projects.
- Add someone to the implementation team who is still in organizational change management.
- Start holding weekly town meetings to keep everyone informed about the major business issues.
- Plan a large digital board that will indicate the number of people waiting for their calls to be serviced.

Now, the improvement can be implemented in hours to six weeks. There just is no reason why a combination of these cannot be implemented within a 30-day period. To accomplish this an individual should be assigned to each of the proposed improvement solutions to develop estimates on costs, risks, and cycle time. These estimates should be presented to the LTM team for comments and to confirm the estimates. Don't feel that the estimates have to be exact—a good approximation will fulfill the purpose. For those proposed improvement solutions that are accepted by the sponsor, additional time will be taken to refine these estimates.

The author cannot overemphasize the importance of developing a cost–benefit analysis that is as accurate as possible. To get analysis that is plus or minus 10% is rare in the time that is allotted in the LTM approach. In many cases, the LTM team only focuses on dollars saved; in reality, they need to estimate the impact for all the three primary improvement targets. We recommend that someone from finance be invited to help with this particular task as they have the credibility and experience in estimating costs. Often, in defining the improvement plan, the LTM team will be divided up into two or three working units, allowing each unit to develop its own approach. In this case, you will need to make two or three independent cost–benefit analyses. In reality, what the LTM team is doing is a value analysis of the proposed improvement approaches.

All too often, the LTM team focuses on the analysis of potential savings without doing an acceptable job at estimating implementation cost. These conditions can lead to a serious error in the implementation approaches that are finally agreed on. Figure 6.12 is a typical example of how three different sets of improvement approaches will have on the organization's performance.

Figure 6.13 is an estimate of the implementation resources estimated to implement each of the three options.

If you look only at *Change 3* in Figure 6.13, it is estimated to have better performance in all but one area. But, when you look at the implementation costs and cycle time in Figure 6.13, it's obvious that *Change 2* provides the most value to the organization if the product will only be produced for three years.

The LTM team will determine the resources required to implement the improvement and the estimated impact that the improvement will have

Performance estimate

	Original process	Change 1	Change 2	Change 3
Effectiveness (quality)	0.2	0.02	0.01	0.009
Efficiency (productivity)	12.9 hrs./cycle	7.5 hrs./cycle	6.3 hrs./cycle	5.3 hrs./cycle
Adaptability measurement	25%	55%	80%	65%
Cycle time	305 hrs.	105 hrs.	105 hrs.	85 hrs.
Cost per cycle	$605	$400	$410	$330

FIGURE 6.12
Performance improvement estimates. (© 2007, Harrington Institute, Inc.)

Implementation estimate

	Change 1	Change 2	Change 3
Cost	$1,300,000	$20,000	$280,000
Implementation cycle time	12 months	2 months	3 months
Probability of success	50%	95%	85%
Major problem	Need more data	Training time	New org. structure

FIGURE 6.13
Implementation resources required estimate. (© 2007, Harrington Institute, Inc.)

upon the product and/or process. For this reason, often a member from the finance department and/or the cost-estimating department are active members of the LTM team. The LTM team members should not spend an excessive amount of time trying to get very accurate estimates related to implementation resources required and improvement gains that will result from the improvement initiative. Typically, boxing this activity into 45 minutes is adequate. Add an additional 15 minutes at the end of the estimating cycle if it looks like it would be fruitful.

Once the high-level suggestions have been defined, the facilitator will walk the team through the steps needed to create an LTM action plan, which has seven key parts. Typically, it can be recorded on one or two sheets of paper when the plan is crisply defined. (See Figure 6.14.) The LTM action plan consists of the following:

- *Improvement opportunity*—This is a short description of the current issue, improvement opportunity, or problem as identified during the workshop.
- *Recommendation*—This defines what action needs to be taken to correct the problem.
- *Action plan*—This is a step-by-step plan that defines how the recommendation will be implemented, who will be responsible for implementing each specific bullet, and the date the activity will be completed.
- *Cost benefits*—This is an analysis of what it will cost to implement the suggestion and what the paybacks will be from the implementation. By dividing the implementation cost into the payback, you get return-on-investment ratio. For example, if it costs $10,000 to

Lean TRIZ action plan		
Improvement opportunity		
Recommendation		
Action plan	Responsible person	Date completed
Cost benefits (Define in terms like cost, processing time, and quality)		
Change sponsor(s)		
Change champion(s)		
Status: Approved ______Declined ______Pending ______Resolution date ________		

FIGURE 6.14
LTM action plan form.

implement the suggestion and the payback was $100,000, the payback would be 10 to 1 return on investment. Of course, benefits are not always dollars saved; it can be processing time, increased customer satisfaction, increased customer loyalty, decreased cycle time, increased market share, extended products' life cycle, or improved quality. In reality, this is just the way value analysis is calculated.

The value analysis is one of the most difficult parts of the LTM project. Often, it is very hard to get agreement to qualify things like what is the benefit for improved customer satisfaction or reducing cycle time. What are the real savings from reducing inventory? In most of these cases, it is a lot more than we would first realize. More advanced organizations have established standards for all of these important savings that will make the improvement estimating as easy as possible. If your organization has not matured to that level, then your best estimate will have to do. In the less refined organizations, things like improved customer satisfaction and improved employee morale are listed as intangible savings.

The LTM champion should at the very beginning of the program contact the financial department to get their input on things like time cost of money, and what part of the overhead cost is variables overhead. (This is the part of overhead that is eliminated when a job is eliminated.) They can also provide other established guidelines that have been standardized throughout the organization that will help the LTM team members in performing the value analysis.

Note: The following list will help clarify some of the terms in Figure 6.14.

The following are some key definitions:

- *Change champion*—This is the individual who is responsible for coordinating the day-to-day implementation of the recommendation(s). The individual who is assigned this responsibility should be a member of the workshop team. The change champion is the one who makes it happen.
- *Change sponsor*—He or she sponsors individual changes by providing resources and measures performance throughout the implementation. The change sponsor is accountable to the LTM sponsor for implementation. During the implementation process, the change champion may face challenges such as conflicting priorities that he or she is unable to resolve. The change sponsor is then responsible for resolving these issues.
- *Status*—This is the status of the recommendation once it has been reviewed by the LTM sponsor. The change sponsor should either approve or decline the change. In rare occasions, he or she may put it in the pending category, but, when it is placed in the pending category, it must have a resolution date when he or she will make a final decision.

Once the facilitator has taken the team through the sequence of preparing an LTM action plan, the team will be divided up into small groups who will complete the LTM action plans. The leader of each of these subgroups should be the change champion for his or her part of LTM action plan.

PHASE III: CONDUCTING THE WORKSHOP

Activity 6: Management Review and Approval

Task 11: Preparing the Sponsor's Presentation

The LTM members have put in a lot of work over the past one or two days, preparing a set of recommended improvement changes. The success of these efforts will be measured by the results of the meeting with the sponsor. Doing a good job in preparing for the meeting is crucial to the final outcome of the project. The team must decide on the agenda for the meeting and who will present each of the agenda items. The team will need to select presenters who can sell the recommendation to the sponsor. Try to select presenters who are used to presenting to upper management and/or who have a very effective presentation style. The presentation will last for one to three hours depending upon the number and complexity of the recommended improvements. At the end of the second day, everyone will be tired and if the presentation is scheduled for Friday or Sunday, everyone will be pushing to go home as soon as possible. It is very important that the presentation is crisp and keeps moving. Keep the number of presenters down to a minimum—no more than two to three per hour. The facilitator should point out to the executive team that they are encouraged to add positive comments during the presentation. We have found it effective to start preparing parts of the presentation during Task 9, as soon as individual approach is evaluated as feasible. This is usually done by assigning one or two team members to break off from the team's activity to prepare that portion of the management presentation. The sponsor will realize that the management presentation may not be as professional looking as he or she is normally observing due to the short time allotted for the workshop.

The sequence of the presentation is also very important. We suggest that it starts with quick hits—the ones that are easy to understand and get approved. These require less discussion and justification to get the meeting started on a positive tone. Other key managers (aside from the LTM project sponsor) who will need to provide resources during the implementation stage should be invited to attend. It may be necessary to go back to the original charter and present it so that everyone is on board.

The LTM agenda for the workshop should be planned so that adequate time is allotted to prepare for the sponsor's meeting. If power points are the standard method for the meeting presentation, the team should prepare power points for the sponsor's review meeting. Often flip charts are used to show how the cost benefits were developed. We find that allowing 10 to 15 minutes for each of the recommendations is about right. Half of the time should be allotted to the presentation and the remainder of the 15 minutes allotted to discussion and to accepting or rejecting the recommendations.

A dry run of the total presentation should be made to the team by the individual who will be doing the presentation. Nothing polishes a presentation better than giving it to a friendly audience. It will also provide fresh input that will add additional value to the recommendation. Again, timeboxing in each recommendation is strongly recommended in order to get through all of the recommendation and complete the presentation on schedule.

A typical sponsor's briefing with new potential change sponsors in attendance can be seen in Table 6.2.

TABLE 6.2

Typical Sponsor's Briefing Agenda

Sponsor's Briefing Meeting	Time
1. Introduction	2 minutes
2. Lean TRIZ charter review	3 minutes
3. Review of final flowchart	5 minutes
4. Describing the workshop process	5 minutes
5. Presenting recommendation number 1	8 minutes
6. Discussion related to recommendation number 1	7 minutes
7. Presentation of recommendation number 2	8 minutes
8. Discussion related to recommendation number 2	7 minutes
9. 45-minute checkpoint	2 minutes
10. Presentation of recommendation number 3	8 minutes
11. Discussion related to recommendation number 3	7 minutes
12. Presentation of recommendation number 4	8 minutes
13. Discussion related to recommendation number 4	7 minutes
14. Presentation of recommendation number 5	8 minutes
15. Discussion related to recommendation number 5	7 minutes
16. Presentation of recommendation number 6	8 minutes
17. Discussion related to recommendation number 6	7 minutes
18. Summary and closure	13 minutes
Total time	120 minutes

You will note that the presentation of the recommendations does not start until 15 minutes into the agenda. The sponsor briefing should start right on time. The presenter should not go back and bring latecomers on board; he or she should volunteer to discuss with them what they have missed after the meeting. There is no real problem in doing this since the first 15 minutes are informational only and do not require the approval of the attendees. It is a good idea to have included in the presentation the LTM team's estimate of the accuracy of the resource requirements and performance improvement projections included as part of the presentation. In many cases, it is good practice to present ideas and improvements that fall outside of the LTM to the sponsors at the end of the presentation. This provides an excellent view of the total concepts that were considered.

The summary should be presented by the team leader, and it should compare the project results of implementing the recommended suggestions to the goals set forth in the LTM workshop charter.

Task 12: Presenting to the Sponsor(s)

The sponsor briefing is designed to present the LTM recommendation to the sponsor(s) and other key people so that the LTM team gets immediate feedback upon their recommendations (accepted, rejected, or pending). The LTM sponsor will always attend these briefings. Other change sponsors will also be invited to attend as they are identified. It may be necessary to put a recommendation in the pending status when a change sponsor can't attend the briefing and his or her representative doesn't have the authority to commit the required resources. The objective of this briefing is to do the following:

- Get change sponsor's support
- Obtain decisions on recommendations
- Get committed resources for the implementation of the approved recommendations

It is important to point out at the very start of the meeting that this is a decision-making meeting, not a problem-solving meeting. The time allotted for the briefing doesn't allow time for other opinions to surface and be discussed unless the recommendation is rejected. In this case, a discussion related to how this recommendation can be modified to make it acceptable is encouraged. A strict set of ground rules should be defined and presented at the beginning of the meeting. Some of them are the following:

- Strict adherence to the agenda and time schedule.
- Only three decisions can be made:
  1. *Accept*—Go ahead with the implementation.
  2. *Reject*—Terminate work on the recommendation.
  3. *Pending until a decision on the recommendation will be made within five days*—All recommendations that are put in the pending category should have an agreed-upon time and a place where the LTM sponsor will make a final decision. We recommend that that decision be made during the presentation(s).

Typical rules that will help keep the meeting on schedule are the following:

- Change champions and change sponsors for all approved recommendations will be assigned at this meeting.
- The facilitator will determine when a discussion should be terminated.
- Determine which sponsor has the ultimate decision.
- There will be no side conversations.
- There will be open dialog.
- All points of view are welcome.
- Only one person should talk at a time.
- Everyone should be respectful of one another.
- The facilitator acts as the gatekeeper.
- No one should leave before the end of the scheduled meeting.

We suggest that the sponsor's briefing be held in the same room that the workshop was conducted in. This provides the sponsor(s) a view of the rainbow of activities that the LTM members went through to come up with their final recommendations. We have witnessed sponsors staying around for hours after the meeting just to understand the LTM process better. Often, they then become real advocates of the LTM program.

It is important that the results are well documented including the individual who is responsible for managing each suggested improvement. The following is a time-boxed agenda for a typical LTM team:

Timeboxing the workshop

Day one agenda

- 8:00–8:20 a.m.—Phase III. Activity 1: Introduction
  - Task 1: Introduction and business overview and getting a sponsor on board.
  - Task 2: Present an overview of Lean TRIZ methodology.
- 8:20–9:45 a.m.—Phase III. Activity 2: Develop the as-is model.
  - Task 3: Review situation questionnaire results and agree on root causes and major contradictions related to the improvement opportunity.
  - Task 4: Refine the process flowchart or review the current product specifications.
  - Task 5: Refine the as-is matrix.
- 9:45–10:00 a.m.—Break
- 10:00–10:45 a.m.—Phase III. Activity 3: Start workshop and define root causes.
  - Task 6: Identify issues, opportunities, problems, and root causes that were not covered in Task 5.
- 10:45–11:00 a.m.—Phase III. Activity 3: Start workshop and define root causes.
  - Task 7: Prioritize issues, opportunities, problems, root causes.
- 11:00–1:30 p.m.—Phase III. Activity 4: Define potential improvement options.
  - Task 8: Define and quantify improvement action for issues, opportunities, problems, root causes and major contradictions.
    - Task 8.A: *For process-related and redesign improvement opportunities*, use brainstorming, streamlined process improvement, Lean, and LP-TRIZ.
    - Task 8.B: For *product design and redesign improvement opportunities*, use brainstorming, value engineering, and TRIZ.
- 1:30–2:30 p.m.—Lunch
- 2:30–3:35 p.m.—Continue activities for Phase III. Activity 4: Define potential improvement options.
  - Task 8: Define and quantify improvement action for issues, opportunities, problems, root causes, and major contradictions.
    - Task 8.A: For *process-related improvement opportunities*, use brainstorming, streamlined process improvement, lean, and LP-TRIZ.

  - Task 8.B: For *product design improvement opportunities*, use brainstorming, value engineering, and TRIZ.
- 3:35–5:00 p.m.—Phase III. Activity 5: Quantify improvement opportunities.
  - Task 9: Prioritize and classify improvement action items.
  - Task 10: Quantifying and qualifying potential improvement solutions.
- 5:00 p.m.—End of workshop day.

Day two agenda

- 8:00–12:00 p.m.—Phase III. Activity 6. Management review and approval.
  - Task 11: Prepare the sponsor's presentation.
- 12:00–1:00 p.m.—Lunch
- 1:00–3:00 p.m.—Phase III. Activity 6. Management review and approval.
  - Task 12: Present to the sponsor.
- 3:00–5:00 p.m.—Phase III. Activity 6. Management review and approval.
  - Task 13: Summarize the results and document them.
- 5:00 p.m.—Workshop over.

At the beginning of the meeting, the ground rules should be presented and understood. Following this, a discussion of how the workshop functions is appropriate. This is followed by presentations of the individual recommendations. Each recommendation is presented, followed by a discussion where a decision is made to accept, reject, or put the decision off for five days. It is imperative that the facilitator keeps the discussion moving and documents all decisions. These decisions should be documented visually to the entire attendees either on a flip chart or on a whiteboard (see list below). The change champions are responsible for documenting the specific discussions related to their ideas.

Documented status during meeting

- Idea number one—Relay all of the departments—Accepted Richard Langley responsible for implementation. Completed within three weeks.
- Idea number two—Redo the overhead lighting—Rejected as return on investment does not justify the change.

- Idea number three—Relocate the drinking fountain to reduce interruptions within the department—Accepted Billy Thompson assigned. Completed within 10 days.
- Idea number four—Remove signoff requirements for second and third level managers, documents will be sampled to ensure they are legitimate—Accepted Tom Jenkins responsible. Completed within five working days.
- Idea number five—Installed a new software tracking system—Idea rejected due to low payback.
- Idea number six—Install a special handling system for rush orders—Accepted Tom Jenkins's responsibility. Completed within six weeks.

Note: Even when an action plan is put on hold, the reason for it being put on hold should be recorded in the comments section.

When the team leader presents the summary statements, he or she will schedule a follow-up meeting to plan for the implementation. Attendees at this implementation meeting should include the team leader, the change sponsors, and the change champions.

Often, a town meeting is scheduled where the people who will be implementing the approved recommendations and the people who will be impacted by these recommendations are invited to attend. We recommend that the town meeting be held within two days after the LTM sponsor briefing. This is a very effective way to get buy-in to the recommendations and to keep the people affected by them informed. Good communication is key to a successful implementation. You can't have too much of it. We have also seen situations when the discussions during the town meeting have pointed out refinements in the recommendations that meant the difference between success and failure.

Task 13: Summarize the Results and Document Them

The workshops should be well documented. All decisions, issues, and suggestions made during the sponsor briefing should be documented. If the workshop output was captured on a computer, then copies of the following should be handed out at the beginning of the sponsors' briefing:

- Final workshop charter
- Final process flowchart

- The payoff matrix
- All LTM action plans

This phase is completed when the team leader turns in a summary that documents the output from the sponsor's briefing.

SUMMARY

This phase requires a high degree of creativity from the LTM team. It covers how the two-day workshop should be managed in order to make it profitable. It covers all the following activities:

- Introducing the LTM team to the methodologies
- Defining root causes
- Collecting data
- Developing best future-state solutions
- Estimate costs to install
- Estimate potential savings
- Presenting the team's recommendations to the sponsor

These are the two days that everyone has been preparing for, and teams will need to stick to the schedule in order to get the desired results. It uses a technique called timeboxing, and you need a very good and skilled facilitator to implement this approach. Using this approach, we have seen a number of occasions where the focused two days resulted in more than \$1 million savings using either the Process Lean TRIZ or the Product Lean TRIZ approach. The members of the LTM team were exposed to TRIZ, Lean, value engineering, business process improvement, and LTMs. Although their depth of training is inadequate for them to perform any one of these methodologies by themselves, it often is the stimulus that gives them the desire to obtain a detailed understanding of one or more of these improvement methodologies. The real key to holding a successful workshop is the amount of planning, data collection, and preparation that the facilitator and the LTM team champion invest in acquiring good, sound data prior to starting the workshop.

The difference between success and failure is the preparation before the workshop.

Real change is not what you do but what happens after you did it.

H. James Harrington

7

Phase IV—Implementing the Change (Recommendations)

The best ideas are wasted effort unless they are put into action.

H. James Harrington

INTRODUCTION

After you have conducted the workshop, you move into Phase IV—Implementing the change (recommendations). (See Figure 7.1.)

The approved workshop's suggested changes must be implemented, or the total project is a failure. The Lean TRIZ methodology (LTM) program can be dropped if the suggested improvements are not implemented and are not effective. This is the quickest way to turn off management and the employees on all future improvement approaches.

The following are the activities that take place during this phase:

- Activity 1: Sustaining the momentum
- Activity 2: Developing an implementation plan
- Activity 3: Organizational change management (OCM)
- Activity 4: Implementing the project plan

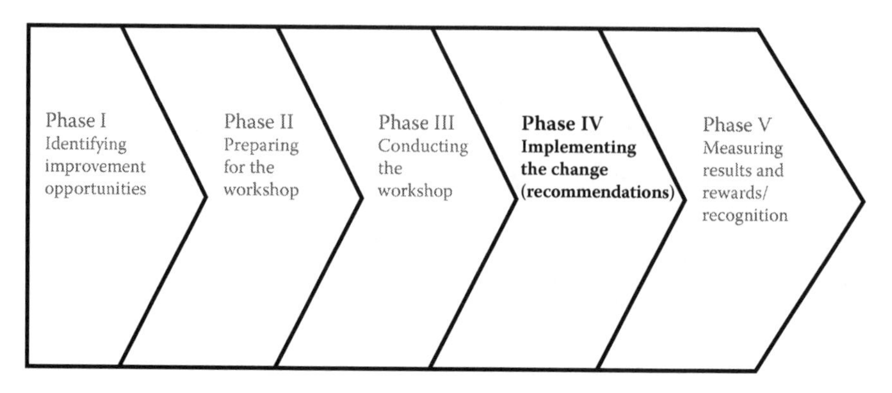

FIGURE 7.1
Phase IV—Implementing the change (recommendations).

ACTIVITY 1: SUSTAINING THE MOMENTUM

Brainstorming problems, creating solutions, and presenting to top management are all fun and exciting things to do, but implementing is just hard work. During the workshop, the attendees should devote all of their time to the project without other interruptions. Anything that came up should be put off to the next day. Their managers had assigned them to do the LTM workshop, and very few things should be of higher priority. They were away from their phones and their computers, so their concentration was not interrupted. At the end of the workshop, they are back in the real world—the phone is interrupting their meetings, priorities are changing daily, if not hourly, etc. The two days at the workshop put them two days behind their normal workload, and now they have to catch up on these things they had put off. In addition, they have accepted some new additional tasks that need to be done within the next 30 days.

The big challenge that the LTM sponsor and the LTM champion face is how to keep the same level of commitment to the project that the team had during the workshop all the way through the implementation phase. To make the problem more complex, sometimes, the people who are responsible for the implementation of the recommendations are not the same ones who attended the workshop. The question is, "How do you sustain the momentum that was developed during the workshop throughout the implementation phase?" There are just too many great ideas that die and are buried in the file cabinet or in the bottom desk drawer because they were not implemented. A good idea is a wasted effort if it is not put to work. Well, this difficult job rests firmly on the shoulders of the LTM champion, the team leader, and the LTM sponsor. Too often, the change champion doesn't have the required experience on how to run a project. This often requires that the change champion receive basic training in project management methodologies.

ACTIVITY 2: DEVELOPING AN IMPLEMENTATION PLAN

To keep the enthusiasm up, the results of the workshop should be communicated to the organization in its newsletter or posted on its bulletin boards. The LTM facilitator should also send a personal letter to each member of the workshop, recognizing his or her contributions and stating how important it is to implement the good ideas.

The team leader and the facilitator should schedule meetings with the change sponsor and change champion combination. During the meeting, the following issues should be discussed:

- Purpose of the meeting
- Review implementation guidelines
- Review goals and responsibilities
- Review the assigned recommendation(s)
- Develop a work breakdown structure
- Review the project management methodology
- Define how results will be measured
- Define the type of people who should be on the implementation team and who they should be
- Define how the needed resources will be obtained

When the team is assigned, the LTM sponsor and the team leader should both attend the first meeting of each implementation team to point out what the team members' roles and responsibilities are and how important it is to complete the assignments in an expeditious manner. They should also point out that the team is part of a total improvement process. It is usually helpful for the team members to know about the other projects that are being implemented at the same time. Once a week, a change champion meeting for the process will be held. This meeting will be chaired by the team leader, and the LTM sponsor will often attend. At this short,

TABLE 7.1

Typical Work Breakdown Structure

Project name:				
Prepared by:				
Start date:	Target finish date:			
Step	Task	Start date	Finish date	
10	Task			
	Status			
20	Task			
	Status			
30	Task			
	Status			

50-minute meeting, each change champion will report the status of his or her activities, keeping all of the key people informed about the progress that has been made.

These changes can't be made in isolation. A change in one part of the process can cause major problems later on in the process. The change teams must be careful not to suboptimize their part of the process at the cost of other parts of the process.

The team leader now has a portfolio of projects to manage that can take a high percentage of his or her time. We find that a simple program like Microsoft Project is an effective way to manage the portfolio of projects. Others believe that it is too complicated for a short, 30-day project and prefer to do it with a simple list. Table 7.1 is a simple way of documenting a work breakdown structure.

ACTIVITY 3: OCM

The team leader and the LTM sponsor have another job that must be addressed with the major changes to the process and/or the product, i.e., there is often a new need for a measurement and feedback system.

We all like to think of ourselves as change masters, but, in truth, we are change bigots. Everyone in the management team supports change. They want to see others change, but, when it comes to the managers themselves changing, they are reluctant to move away from past experiences that have proven successful for them. If an organization is going to change, top management must be the first to do so.

Definition: Organizational change management is a comprehensive set of structured procedures for the decision-making, planning, execution, and evaluation phases of the change process.

We can think of change as a real process; just like any of the processes that go on within the organization, it is the process of moving from the present state to the as is through a transitional period that is extremely disruptive to the organization: to a future desired state that somebody believes is better than the current state.

Definition: Change is a condition that disrupts the current state.

Definition: Transition state is a point in the change process in which people break away from the status quo. They no longer behave as they have done in the past, yet they still haven't thoroughly established the way of operating.

The transition state begins when the solution disrupts the individual's expectations, and they start to change the way they work.

Definition: Future state is the point at which change initiatives are in place and integrated into the behavioral patterns that are required to make the change successful.

It doesn't matter whether you're a big organization or a small organization—your employees will resist change. Future state is when the change goals and objectives have been achieved.

People are very control oriented. They are the happiest and most comfortable when they know what's going to happen and their expectations are fulfilled. Keep this in mind. For example, if I came home from a long trip at 11:30 p.m. and found all the doors unlocked and the lights on but my wife wasn't home, that would be a change for me. It's customary for her to greet me at the door with a kiss. My expectations would have been disrupted. People resist change because they are dissatisfied about the disruption to the current status as much or more than they are afraid of the change. How would I react to this disruption in expectations? I would call my son to find out if he knew where his mother was. I'd wake up my wife's friends to see if they knew anything about her whereabouts. I would call the local hospitals. I would be upset, worried, and unhappy.

When change occurs and expectations are not met, the following four C's come into play:

1. Employees feel less *competent*.
2. Employees feel less *comfortable*.

3. Employees doubt their *confidence.*
4. Employees feel that they aren't competent enough to handle the unknown that comes with change.

Change makes people feel uncomfortable because they are entering a world they have not experienced before. Change in the work environment causes people to lose their confidence. Before the change, they knew their job better than anyone else, but now they will need to start to learn over again. Change causes people to feel that they have lost control over their lives and activities. From an individual standpoint, the people who are making the change are controlling their destiny. The person has lost control over his or her own life. With disrupted emotions within the organization, it quickly heads in a negative direction. Stress levels go up very quickly because people start to worry about what will happen to themselves and their friends. Productivity drops off as people make time to discuss what's going to happen to them and start to question whether they are doing the right thing. The organization becomes unstable as people start to react more slowly to the proposed change.

The focus on OCM implementation methods is on the transition from the present state to the future state. The journey of this transition can be long and perilous, and, if not properly managed and without using appropriate strategy and techniques, the results can be disastrous.

OCM is a process that has over 25 different tools associated with it. It is not our intention to present in this book the total OCM concept. We have decided to discuss only one of the many tools—pain management.

Pain Management

One of the main issues in any change project is achieving *informed commitment* at the beginning. You can apply a basic formula that addresses the perceived cost of change versus the perceived cost of maintaining the status quo. (See Figure 7.2.) In this case, when we talk about cost, it isn't just dollars and cents; more importantly, it is the emotional stress that the individual affected by the change will be subjected to. As long as people perceive the change as being more costly than maintaining the status quo, it is extremely unlikely they will allow the change to be successful. You need to increase people's perception of the high cost of maintaining the

FIGURE 7.2
As-is pain versus future-state pain.

status quo and decrease their understanding of the pain related to going through the transformation plus the future state pain so that people recognize that, even though the change may be expensive and frightening, maintaining the status quo is no longer viable. And it is, in fact, more costly. This is the process called *pain management*. The LTM team needs to build enough commitment in the people who will be impacted by the change so that they will sustain the output from the change process. Someone on the LTM team should be assigned to minimize resistance to the proposed changes by the individuals who will be affected by the change.

ACTIVITY 4: IMPLEMENTING THE PROJECT PLAN

Implementation—what can you say about it? It's a matter of selecting the right people, having an integrated plan that ties it all together, having people that live up to commitments, and having adequate follow-up to ensure the plan is being implemented as designed.

> A poor plan implemented effectively yields more results than a good plan implemented poorly.
>
> **H. James Harrington**

SUMMARY

No matter how good the plan is or how good the improvement strategy is—the project is a failure if it is not implemented. Unfortunately, most of our professional people get personal satisfaction out of developing corrective action and strategy. The hard work of implementation is often transferred to lower-level people who were not involved in creating the initial strategy. This is a major mistake that many organizations make because management seems to be more interested in developing comprehensive plans than in implementation.

> Innovation, implementation, and integration are all required to get maximum results.
>
> **H. James Harrington**

8

Phase V—Measuring Results and Rewards/Recognition

INTRODUCTION

The final phase is Phase V—Measuring results and rewards/recognition. (See Figure 8.1.)

A Lean TRIZ methodology (LTM) project is not completed until the results of the activity can be measured and the appropriate recognition has been given to the individuals involved. All too often, the individuals who worked on the project are reassigned to other projects before enough data can be collected to effectively measure the impact of the project. As a result, the individuals and teams that work on the project are not rewarded for the additional effort they put in to make the project successful. To keep this from happening, as a standard procedure, the following three activities need to be implemented:

1. *Activity 1: Measurements are key to improving.*

   Effective measurement systems collect enough information to adequately measure the impact the project has on the organization. These key measurements must include the cost of designing a change, implementing the change, and measuring results.
2. *Activity 2: Group recognition.*

   There needs to be established a recognition system that acknowledges different levels of performance for groups related to a specific project.
3. *Activity 3: Private recognition.*

   There needs to be established a recognition system that provides awards based upon the individual's performance on the specific project.

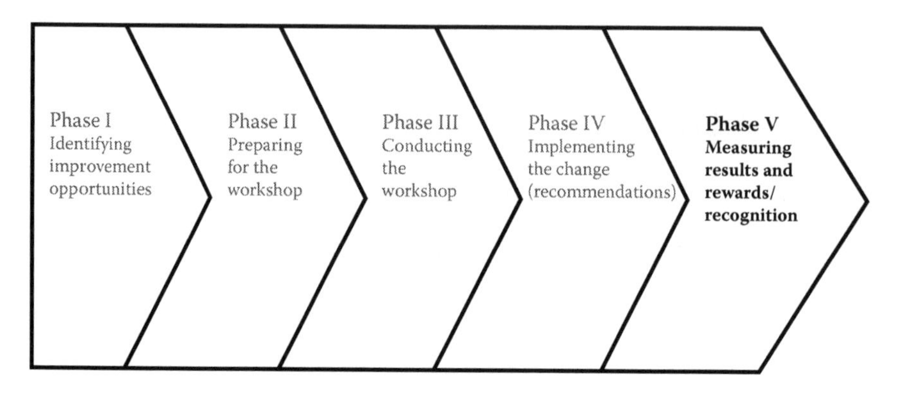

FIGURE 8.1
Phase V—Measuring results and rewards/recognition.

It is very important that the standard procedure that is used by all parts of management be developed and implemented. Without this standard, some areas will recognize performance that is substandard, and others will have much higher requirements.

ACTIVITY 1: MEASUREMENTS ARE KEY TO IMPROVING

Dr. Charles Coonradt, a management consultant and founder of The Game of Work, believes that we should make business more like a game. He points out that the same workers who require special clothing to work in the cold storage section of a plant for only 20 minutes, with a 10-minute break to get warmed up, will spend an entire day ice fishing on a frozen lake. When the air conditioning breaks down, and the temperature reaches 85°, the office is closed. However, on the way home, employees will stop and play a round of golf in the same heat—and pay to do it.

Why is this? Sports succeed in exciting people because they have rules, measurements, and rewards. Let's apply these principles to business:

- *Rules*

  Every sport has rules that govern the game. People know them and are penalized when they do not live up to them (e.g., a 15-yard penalty for clipping another player in football). Business, too, has its rules. They are the procedures and job descriptions. Business also has referees—the quality system auditors. When you do not play by the rules, you should be penalized.

- *Rewards*

  Sports enthusiasts get their rewards from trying to improve their games. The professionals see big dollars rolling in, and the amateurs win trophies, but all are tied into a measurement system. In business, we should all be pros, or we should not have our jobs. Sure, it is nice to get the trophies, plaques, and dinners when we excel, but we should also receive financial awards. Our pay should be tied directly to our personal measurement system.
- *Measurements*

  How much fun would it be to play tennis and not know how well we are doing? How popular do you think golf would be if the only feedback you received is that, of the 200 people who played golf last Sunday, the average score was 93, and you weren't told what your score was? Yes—measurements are critical to maintain interest in an activity, particularly when you want to improve. Measure both individual and team performance. A baseball team, not an individual, wins or loses the game. Apply team measurements to small groups (3–10 people).

Measurement is important for improvement for several reasons:

- It focuses attention on factors contributing to achieving the organization's mission.
- It shows how effectively we use our resources.
- It assists in setting goals and monitoring trends.
- It provides the input for analyzing the root causes and sources of errors.
- It identifies opportunities for ongoing improvement.
- It gives employees a sense of accomplishment.
- It provides a means of knowing whether you are winning or losing.
- It helps monitor progress.

Service organizations tend to think, or maybe even believe, that they cannot be measured. Consequently, in the past, we measured only products and ignored service organizations. This happened for several reasons:

- Traditional measures of input and output could not be applied easily to service organizations.
- White-collar workers believed that their work was varied and unique and could not be measured.

- In the 1950s and 1960s, product cost (material and labor) constituted a significant portion of total cost. Hence, management was preoccupied with measuring and managing it. Business process cost was a smaller element of cost and was therefore ignored.
- Measurement trends are changing,
    - From product measurement to process and service measurement
    - From managing profits to managing assets
    - From meeting targets to continuous improvement
    - From quantity measurements to measurements focusing on effectiveness, efficiency, and adaptability
    - From measurements based on engineering and business specifications to measurements based on internal and external customer expectations
    - From a focus on the individual to a focus on the process (In the past, it was assumed that individuals could control all results; now, we accept that it is the process that should be measured.)
    - From a top–down process dictator approach to a team approach

Obviously, all business processes and service organizations can, and should, be measured. They should be managed and measured in the same way manufacturing processes are measured and controlled.

A rewards and recognition system that reinforces the desired behavior of the LTM members needs to be in place. People need to be accepted by other human beings and recognized for their efforts. Love is a form of recognition. It indicates that an individual is different and special. Your salary is recognition of what your time is worth to your company. A baby cries to get attention, to be recognized. The whole competitive society we live in is driven by people striving to be recognized. Often, recognition is simply having someone else acknowledge they did something well. It is something everyone wants, needs, and strives to obtain. Studies have shown that people classify recognition as one of the things they value most.

Ingredients of a Company Recognition Process

A good recognition process has six major objectives:

1. To provide recognition to employees who make unusual contributions to the company and to stimulate additional effort for further improvement.

2. To show the company's appreciation for superior performance.
3. Having an effective communication system that highlights the individuals who were recognized, which motivates other employees to excel.
4. To provide many ways to recognize employees for their efforts and stimulate management creativity in the recognition process. Management must understand that variation enhances the impact.
5. To improve morale through the proper use of recognition.
6. To reinforce behavioral patterns that management would like to see continued.

Why does recognition matter? George Blomgren, the president of Organizational Psychologists, puts it this way: "Recognition lets people see themselves in a winning identity role. There's a universal need for recognition and most people are starved for it."

A National Science Foundation (NSF) study made the same point. "The key to having workers who are both satisfied and productive is motivation, that is, arousing and maintaining the will to work effectively—having workers who are productive not because they are coerced but because they are committed." The NSF study continues, "Of all the factors which help to create highly motivated/highly satisfied workers, the principal one appears to be that effective performance be recognized and rewarded—in whatever terms are meaningful to the individual, be it financial or psychological or both."

There are five major types of recognition:

1. Financial compensation
2. Monetary awards
3. Personal public recognition
4. Group public recognition
5. Private recognition

Although all of these can be used with LTM, we will just discuss the last two—(1) group public recognition and (2) private recognition.

ACTIVITY 2: GROUP RECOGNITION

A recognition of the group makes the group feel that it is a winner, and the members of the group get a sense of belonging that leads to increased company loyalty. Again, there are an unlimited number of ways for management to recognize a group for its contributions. Typical ways are the following:

1. Articles about the group's improvement are posted in the company's newsletter, accompanied by photographs of the group.
2. Department luncheons are held to recognize specific accomplishments.
3. Family recognition picnics are organized.
4. Progress presentations are made to upper management.
5. Luncheons with upper management are scheduled.
6. Group attendance at technical conferences is encouraged.
7. Cake and coffee are offered at a group meeting, paid for by the company.
8. Department improvement plaques are distributed.
9. Top management attend group meetings to say thanks for a job well done.
10. Group mementos (pen sets, calculators, product models, etc.) are given out.

ACTIVITY 3: PRIVATE RECOGNITION

Of all the recognition categories, this is one of the most important because it directly relates to the interface between management and the employee. The one-on-one interface is very important in stimulating improvement and keeping morale high.

There are many unpleasant jobs that have to be performed, that can't be automated, that can't be ignored, and that have no prestige associated with them. However, if these jobs weren't performed, an organization could not function effectively. What makes these unbearable jobs bearable for the employee is a manager who appreciates the individual's contribution and lets the employee know that the effort being put forth is appreciated. This is the type of manager who makes a good job great and a great job fantastic. Such a manager always seems to *luck out* and have the best employees. It's that department that always seems to get out the work on schedule, seemingly exerting no effort, and, when the company has a problem, it's that area that is first to step forward and volunteer to help. This area's absenteeism is down and productivity is up. Why? Because the manager remembers asking an employee to do something and follows up to see that it was accomplished, not looking over the employee's shoulder but rather showing that the assignment was important and that the effort the employee invested is noticed.

A lot of managers feel strange telling employees they are doing a good job, and, frequently, the employee has a hard time accepting compliments and reacts with comments like "Knock off that bull!" or "Don't give me compliments, give me money!" But comments like that do not mean that they don't need the manager's appreciation. So don't let it prevent you from expressing your honest appreciation for a task well done. Employees need encouragement and need to have their good acts reinforced through management appreciation. A sincere pat on the back at the right time is much better and more effective that a swift kick in the pants at any time.

Typical ways that private recognition is provided to an employee are the following:

1. A simple, honest thank-you for a job well done, given immediately after the task has been completed.
2. A letter sent to the employee's house by his or her manager or upper manager, thanking him or her for his specific contribution.
3. Personal notes on letters or reports, complimenting the originator on content or layout.
4. Sending birthday cards and work anniversary cards to an employee's house, thanking the employee for the contributions that were made over the past year, not with general statements but with specific examples that let the employee know that management knows that the employee is there and what he or she is doing. The employee may be a little fish in a big pond, but he or she is the king or queen of the family. One of the best ways to boost morale and productivity is by letting the family know how important he or she is to the company.
5. The performance evaluation that takes place every three months is an ideal time to give private feedback to the employee about accomplishments. It should not be the first time you will have expressed your appreciation, but it should be used to reinforce the favorable work patterns and summarize employee accomplishments. The most basic rule of performance evaluation is *no surprises*.

SUMMARY

Establishing a good measurement system that is directed at measuring the effectiveness of the proposed changes is an essential part of the improvement process. Without these measurements, management doesn't know if the program should be extended or dropped. Make sure that the measurement system reflects a before-and-after correlation. Management is looking for an improvement, not the current level of performance. One mistake that is often made by an LTM team is not measuring the process and/or product performance prior to implementing the changes. Equally important is being able to identify individuals or teams who perform in an outstanding manner. Many very good employees have been turned off and stopped giving the extra effort because they don't receive recognition for what they are accomplishing. We've seen a number of employees whose work effort dropped off, and, when we pointed that out to them, they come back with statements like this: "I'm doing as much as the other people out there. Why do you expect more from me? When I work hard and get the job done ahead of time, I just get more work assignments. People around here who are goofing off and just meet and/or miss the schedule get paid as much as I do."

We are closing this book by emphasizing that a job is not done until all the paperwork is closed out, and those individuals or groups who performed in an outstanding manner are recognized and rewarded for their contributions.

> A pat on the back is more effective than a swift kick in the pants.
>
> **H. James Harrington**

Appendix A—TRIZ Body of Knowledge

The following is the TRIZ Body of Knowledge, as prepared by Simon Litvin, Vladimir Petrov, Mikhail Rubin, and Victor Fey in *TRIZ Body of Knowledge* dated 2007. It is also the body of knowledge that is used by the Altshuller Institute.

- *40 Inventive Principles*: The 40 inventive principles that form a core part of the TRIZ methodology invented by Genrich Saulovich Altshuller. These are 40 tools used to overcome technical contradictions. Each is a generic suggestion for performing an action to, and within, a technical system. For example, principle 1 (segmentation) suggests finding a way to separate two elements of a technical system into many smaller interconnected elements.
- *System of innovative standards*: This is a collection of 76 innovative standards, which includes the five classes of innovative standard categories in accordance with the type of innovation problem the innovation standard can deal with:
  1. Complications of subfields (su-fields)
  2. Evaluation of su-fields
  3. Transition to micro and macro levels
  4. Measurement and detection of substance levels
  5. Application of innovative standard
- *Subfield*: (1) A subset of a mathematical field that is itself a field. (2) A subdivision of a field.

BASIC TRIZ CONCEPTS, COMPONENTS, AND TOOLS

1. **Foundational concepts**
   1.1. Dialectics as a philosophical foundation of TRIZ [1]
   1.2. Directional evolution of technological systems [2]

1.3. Technological system [3]
1.4. Functions [4]
1.5. Ideal technological system [5]
1.6. Substance, field, su-field [6]. Su-field resources [7]
1.7. Reflectivity principle [8]
1.8. Ideal substance [9]
1.9. Ideal final result [10]
1.10. Inventive situation. Inventive problem [11]
1.11. Levels of inventions [5]
1.12. Contradictions: administrative, engineering, and physical [12]
1.13. System operator. Multiscreen scheme of talented thinking [13]

2. Trends (laws) and subtrends (lines) of technological system evolution [14]

2.1. Trend of increasing degree of ideality [14]
 2.1.1. Mechanisms of increasing the ideality of technological systems [15]
2.2. Trend of nonuniform evolution of subsystems [14]
2.3. Trend of completeness of system parts [14]
 2.3.1. Subtrend of elimination of human involvement [16]
2.4. Trend of *energy conductivity* of systems [13]
2.5. Trend of harmonization of rhythms [13]
 2.5.1. Subtrends of chronokinematics [17]
2.6. Trend of transition to supersystems [14]
 2.6.1. Subtrend of transition from mono- to bi- and polysystems [18]
 2.6.2. Subtrend of increasing structurization of voids [19]
 2.6.3. Mechanisms of convolution (trimming) of technological systems. Coefficient of convolution [20]
 2.6.4. Subtrend of deployment—convolution [21]
 2.6.5. Trimming of technological systems [22]
 2.6.6. Integration of alternative systems [22]
2.7. Trend of increasing dynamism [23]
 2.7.1. Lines of increasing dynamism [24]
2.8. Trend of increasing su-field interactions [25]
 2.8.1. Lines of evolution of su-fields [26]
2.9. Trend of transition from macro to micro levels [27]
2.10. Trend of matching–mismatching (coordination–noncoordination) [21]
2.11. The general pattern of engineering systems evolution [28]

3. Algorithm for Inventive Problem-Solving (ARIZ)

3.1. ARIZ—a program for inventive problem-solving by identifying and resolving contradictions [29]

3.2. Main line for solving ARIZ problems and ARIZ logic [30]

3.3. Structure and basic notions of ARIZ-85C [31]

3.3.1. Problems–analogs [32]

4. Su-field analysis

4.1. Basic concepts and rules [6]

4.2. Standards for inventive problem-solving [32]

4.3. Structure of the system of standards. System of 76 standards [33]

4.3.1. Standards for system modification [33]

4.3.2. Standards for system measuring and detection [33]

4.3.3. Standards for application of the standards [33]

5. Techniques for resolving contradictions

5.1. Techniques for resolving engineering contradictions (inventive principles)

5.1.1. 40 main inventive principles [34]

5.1.2. 10 additional inventive principles [35]

5.1.3. Duality *principle–antiprinciple* [36]

5.1.4. The Contradiction Matrix [37]

5.1.5. Typical diagrams of engineering contradictions [38]

5.2. Techniques for resolving physical contradictions

5.2.1. Separation principles [39]

5.2.2. Using the separation principles at macro and micro levels [40]

6. Scientific effects

6.1. The concept of database of effects [41]

6.2. Physical effects [42]

6.3. Chemical effects [43]

6.4. Geometrical effects [44]

7. System analysis methods

7.1. Methods to search and formulate inventive problems [45]

7.2. Flow analysis [45]

7.3. Trimming (ideal functional modeling) [45]

7.4. Cause–effect analysis. Formulation of key problems [45]

7.5. Component and structural analysis [45]

7.6. Diagnostic analysis [45]

7.7. Evolutionary analysis [45]

7.8. Function analysis [45]
7.9. Integration of alternative systems [45]
7.10. Failure-anticipation analysis [46]
7.11. Supereffect identification (system improvement without solving problems) [47]

Appendix B—Glossary

39 Characteristics of a Technical System: These are the 39 Engineering Parameters for Expressing Technical Contradictions defined in the late 1960s.

40 Innovative Principles: This is a collection of 40 innovative principles designed by the founder, Genrich Altshuller, on the basis of extensive studies of information on numerous inventions. The 40 Innovative Principles can be used in combination with the Contradiction Matrix (also known as the Altshuller Matrix).

40 TRIZ Principles: These are 40 one- or two-word statements that describe approaches to resolving technical conflict (problems and/or contradictions) that were defined by Genrich Altshuller based on his study of over 200,000 patents. These 40 TRIZ Principles have a twofold purpose:

1. Within each principle resides guidance on how to conceptually or actually change a specific situation or system in order to get rid of a problem.
2. The 40 principles also train users in analogical thinking, which is to see the principles as a set of patterns of inventions or 40 TRIZ Principles.

Champion: This individual is the person who is in charge of the Lean TRIZ methodology (LTM) program for the organization. He or she should have a good understanding of classical TRIZ, Streamlined Process Improvement, and value engineering. We recommend that he or she be a certified TRIZ practitioner, a business process improvement practitioner, and a Lean Six-Sigma Black Belt. Individuals who are certified at the Associate or Green Belt levels usually do not have the experience required to handle an LTM program. The champion is the individual who is assigned the responsibilities to maintain an LTM program throughout the organization.

Change: This is a condition that disrupts the current state.

Change champion: This is the individual who is responsible for coordinating the day-to-day implementation of the recommendation(s). The individual who is assigned this responsibility should be a member of the workshop team. The change champion is the one who makes it happen.

Change sponsor: He or she sponsors individual changes by providing resources and measures performance throughout the implementation. The change sponsor is accountable to the LTM sponsor for implementation. During the implementation process, the change champion may face challenges such as conflicting priorities that he or she is unable to resolve. The change sponsor is then responsible for resolving these issues. The change sponsor may need to be assigned to the LTM sponsor during the presentations or immediately following the meeting.

Contradiction: This is defined as a situation that emerges when two opposing demands have to be met in order to provide the results required. A contradiction is a major obstacle in solving an inventive problem. It is often used as an abstract inventive problem model in a number of TRIZ tools. Three types of contradictions are currently used:

1. Administrative
2. Engineering
3. Physical

Coordinator: In big organizations, when they get many LTM projects going on at the same time, the LTM champion may need someone to help manage the portfolio of projects. This individual is referred to as the LTM coordinator and should have an understanding of the LTM.

Facilitator: This is the individual who has a detailed understanding of the LTM, how to use its many tools, and how to manage a team meeting. He or she helps the project LTM team leader keep the workshop on schedule so that it meets the desired goals. It is helpful if this individual is a certified classical TRIZ practitioner and has a good understanding of classical TRIZ, organizational change management, and Streamlined Process Improvement. Usually, he or she is assigned to two to four LTM projects. He or she will advise the LTM leader regarding the proper tools they should use under the present situation to get maximum results.

Five degrees of design complexity: This is a way of grouping new inventions into five categories that reflect the complexity of the thought pattern that goes into each of the five categories.

- Level 1 inventions are obvious and apparent solutions involving well-known methods and knowledge requiring no new invention of any consequence.
- Level 2 inventions constitute minor nonobvious improvements to a system, using methods known within the domain of discourse but applied in a new way.
- Level 3 inventions involve fundamental improvements to a system involving methods known outside of the domain. This involves applying an idea to the domain that has never been used in the domain previously.
- Level 4 inventions entail the development of an entirely new operating principle and represent radical changes.
- Level 5 inventions represent a rare scientific discovery or the pioneering of a totally new industry altogether.

Five major driving objectives: These are the five major headings that each of the 41 process improvement opportunities relate to. They are the following:

1. Cycle time–reduction improvement opportunities
2. Cost-reduction improvement opportunities
3. People-related improvement opportunities
4. Poor results/reliability or quality improvement opportunities
5. Method improvement opportunities

Future state: This is the point at which change initiatives are in place and integrated into the behavioral patterns that are required to make the change successful. It doesn't matter whether you're a big organization or a small organization—your employees will resist change. Future state is when the change goals and objectives have been achieved.

Lean Process TRIZ (LP-TRIZ): This is a tool that is used most often to improve processes. It uses a greatly simplified version of the TRIZ Contradiction Matrix.

Lean TRIZ methodology: This is an improvement methodology that is designed to bring about rapid improvements/changes to products and processes by defining and implementing the changes that can be quickly identified and easily implemented, thereby reducing the cost and time to bring about improvement and change.

Lines of evolution: These describe in greater detail typical sequences of stages (positions on a line) that a system follows in a specific pattern of evolution in the process of its natural progress. Once these positions are known, the system's current position on a line can be identified, and the possibility of transitioning to the next position can be assessed, for example, become flexible or use micro-level properties of materials utilized. Lines of evolution are grouped under one or more of the patterns of evolution that they support.

LP-TRIZ 41 × 3 Improvement Matrix (LP-TRIZ Matrix): It is a column-by-column analysis of each of the 41 process improvement opportunities evaluating their performance to each of the three primary improvement targets (cost reductions, productivity improvements, and quality/reliability improvements). All improvement activities are directed at improving one or more of these three primary improvement targets.

LP-TRIZ's 41 process improvement approaches: These are 41 activities that are most often taken to bring about improvement in process design and redesign. Each of the 41 activities has a positive impact upon one or more of the three primary improvement targets. They are sometimes called the *41 process design/redesign improvement opportunities.*

Matrix of System Constraints: It makes use of the 39 Engineering Parameters for Expressing Technical Constraints and 40 TRIZ Principles—both the vertical and horizontal that are made up of the 39 Engineering Parameters for Expressing Technical Constraints. Vertical terms are used to pick an action that you want to consider using to improve the item we are working on. The horizontal axis is used to identify negative things that could happen based upon the proposed solution the team is looking at. At the junction of the horizontal and vertical columns, up to four of the TRIZ principles are listed that are most likely to offset the negative situation based upon past experience.

Operators: They are aligned with each line of evolution in a group of possible corrective actions called *operators* that can be used to correct the problem or improve the product or process. An operator is a little nugget of wisdom (recommendation, suggestion) on changes to the system designed to trigger you into thinking how to solve the problem or to improve the process/product under evaluation. Operators also serve as a means to change the system position on the relevant lines of evolution (for example, by suggesting employing a hinge or using special physical effects).

Organizational change management (OCM): This is a comprehensive set of structured procedures for the decision-making, planning, execution, and evaluation phases of the change process. We can think of change as a process, just like any of the processes that go on within the organization. It is the process of moving from the present state, through a transitional period that is extremely disruptive to the organization, to a future desired state.

Patterns of evolution: They are a set of terms or statements that define trends that have strong, historically recurring tendencies in the development/evolution of man-made systems.

Situation questionnaire: This is a document used to collect information related to the improvement opportunity. It is usually a set of questions that the organization feels is necessary to define the root cause of a problem or opportunity. It provides a dual purpose—it provides (1) a checklist of the information necessary to define a root cause and (2) the collected data related to the proposed questions.

Sponsor(s): This is the individual who approves an individual LTM project and decides which changes are implemented. This is a manager who is at a high-enough level that he or she can authorize changes to the process or product without higher-level management approval. The sponsor will attend the last part of the workshop where the LTM team members present their recommended changes to him or her. The sponsor is committed to making a decision at this time on each suggestion. The sponsor must either accept or reject the proposal; he or she does not have the option to put it off to a later date. This firm, on-the-spot, decision-making process is a critical part of the LTM process.

Status: This is the status of the recommendation once it has been reviewed by the LTM sponsor. The change sponsor should either approve or decline the change. In rare occasions, he or she may put it in the pending category, but, when it is placed in the pending category, it must have a resolution date when he or she will make a final decision.

System of innovative standards: This is a collection of 76 innovative standards that includes the five classes of innovative standard categories in accordance with the type of innovation problem the innovation standard can deal with:

1. Complications and the complications of su-fields
2. Evaluation of su-fields
3. Transition to micro and macro levels
4. Measurement and detection of su-fields
5. Application of innovative standard

Team leader: This is an individual who is assigned to lead an individual LTM team. He or she schedules meetings, manages the meeting, and follows up on specific assignments.

Team members: These are the individuals who attend the LTM workshop and create the future-state solutions.

Three primary improvement targets (3PIT): These are the three primary reasons that organizations design or redesign current processes. The 3PIT are the following:

1. Cost reductions
2. Productivity improvements
3. Quality/reliability improvements

Timeboxing: This is a way to manage an individual's or group's time. Timeboxing allocates a fixed time period, called a time box, to each planned activity. Several project management approaches use timeboxing. It is also utilized for individual use to address personal tasks in a smaller time frame. It often involves having deliverables and deadlines, which will improve the productivity of the user.

Transition state: This is a point in the change process in which people break away from the status quo. They no longer behave as they have done in the past, yet they still haven't thoroughly established the way of operating. The transition state begins when the solution disrupts the individual expectations and they start to change the way they work.

TRIZ Contradiction Matrix: This is a 39 × 39 matrix. The vertical column lists all of the 39 characteristics of a technical system. The horizontal top line also lists the 39 characteristics of a technical system. The horizontal columns refer to typical undesired results. Each matrix cell points to principles that have been most frequently used in patents in order to resolve the contradiction.

TRIZ methodology: It is *a problem-solving, analysis, and forecasting tool derived from the study of patterns of invention in the global patent literature*. It was developed by the Soviet inventor and science-fiction author Genrich Altshuller and his colleagues in 1946. In English, the name is typically rendered as the *theory of inventive problem-solving*.

Value: This is the comparison of function to cost. Value can therefore be increased by improving function or reducing costs.

$$\text{Value analysis} = \frac{\text{Function}}{\text{Cost}}$$

Value analysis (VA): This is a systematic and organized decision-making process that evaluates your product and processes from a number of unique perspectives.

Value analysis/value engineering: This is a systematic and organized decision-making process that evaluates your product and processes from a number of unique perspectives. VE provides its user with a number of potential improvement concepts that the user utilizes to stimulate his or her creativity. It is an effective tool in identifying potential design improvements and minimizing the risk of project failure. It serves as a knowledge base of past best practices within and outside of the organization.

Value engineering (VE): This is a systematic method to improve the *value* of goods or products and services by using an examination of function. VE provides its user with a number of potential improvement concepts that the user utilizes to stimulate his or her creativity.

Value engineering project: This is not just a collection of studies and best practices; it also includes well-established practices and policies so that they could be integrated into the product development cycle and the improvement projects. VE and VA go hand in hand. They are a systematic and organized procedural decision-making process. They are designed to create more value to the organization's stakeholders than was previously provided.

Index

Page numbers with f refers to figures.

O

P

9781138216778